The British
Ordnance Department
and Canada's
Canals 1815-1855

THE
BRITISH
ORDNANCE
DEPARTMENT
AND
CANADA'S
CANALS
1815-1855

GEORGE RAUDZENS

Wilfrid Laurier University Press

Canadian Cataloguing in Publication Data

Raudzens, George, 1938-
    The British Ordnance Department and Canada's
canals, 1815-1855

Based on the author's thesis, Yale, 1970.
Bibliography: p.
Includes index.
ISBN 0-88920-071-8

1. Great Britain — Colonies — America — Adminis-
tration.   2. Great Britain. Army. Ordnance
Dept. — History.   3. Canada — Economic conditions —
To 1867.*   4. Canals — Canada — History.   I. Title.

JV1063.R39   325'.342'0971   C79-094402-2

Copyright © 1979

Wilfrid Laurier University Press
Waterloo, Ontario, Canada   N2L 3C5

79 80 81 82 4 3 2 1

Contents

Maps

Preface

Britain's military contributions to the shaping of Canada have long been recognized. The army did more than safeguard the frontiers. As a number of writers have noted, soldiers promoted settlements, stimulated the economy, improved social amenities, and provided a variety of services helpful to pioneer communities. But, with the exception of Charles P. Stacey's work on the relationship between the army and responsible government, the small quantity of writing about such influences is cursory and inadequately documented. An examination of the sources clarifies the nature and magnitude of military aid to the Canadian provinces; it also shows that the most important contributor was an organization as yet virtually unnoticed, the Ordnance Department. The Rideau Canal project and the activities of Colonel John By in turn were the major Ordnance preoccupations. Much has been written about both, but until now no one has identified the importance of the department responsible for them. John By has become more myth than man; in fact he was a most troublesome servant to the Ordnance. When the Canal and the Colonel are placed in their proper context they explain a great deal about the reasons for the kind of effect the Ordnance and the army generally exerted on Canada.

The department is interesting in itself because of the position it occupied within the imperial government. After centuries of growth, by the mid-1820s it had become a body with substantial administrative autonomy. Staffed by both soldiers and civilians, it controlled all military property, arms, equipment, and fortification work, and supplied both the armed forces and civilian departments of state with a wide range of engineering, technical, and scientific services. A full history remains to be written, but the Ordnance experience in Canada is both sufficiently complex and representative to warrant special study. A brick-by-brick chronicle of the works constructed by Ordnance men, though a useful undertaking, is not required here. Details of the building of fortresses such as the Quebec Citadel and Fort Henry, only partly executed by the department, are already known in outline and are better placed in technical studies of military architecture. The Ordnance impact on Canada is better gauged by an analysis of its workings as an organization, of its structure, objects, achievements, and failures. In short, this is primarily an administrative history of one branch of the imperial government which operated in Britain's most important colonies of settlement.

While the story of the Ordnance emerged only after the examination of all available British military records relating to British North America between 1815 and 1855, the opportunity for this research was provided by Yale University, which supported me financially as a post-graduate student, provided funds for study in Britain and Canada, and sent microfilm to me in Australia. The basis of this book is my Ph.D. dissertation, completed in 1970 under the direction of Professor Robin W. Winks. I am particularly grateful for his hard work in correcting my errors and his unfailing encouragement. For expert advice I thank Professor Charles P. Stacey, Professor Sydney F. Wise, Professor Arthur Keppel-Jones, Dr. Richard Glover, Professor Richard Preston, and the late Dr. J. Mackay Hitsman. The staff of every archives and library concerned provided courteous assistance and Dr. K. L. Davies and Mr. J. K. Bates of the Scottish Record Office, Mr. Jack Martin of the Royal Military College Library at Kingston, and Mr. Francis Needham, secretary to the present Duke of Wellington, were especially helpful. Colonel E. E. N. Sandeman, librarian of the Royal Engineers Corps Library at Chatham, offered full co-operation and in addition extended the hospitality of the Royal Engineers Mess on more than one occasion. Macquarie University supplied the funds and facilities for revision and preparation of the finished manuscript. My colleagues Miss Alison Affleck, Professor Bruce E. Mansfield, Dr. Robert J. Durham, Jr., Dr. P. N. Lamb, Dr. John Hammond Moore, and Dr. T. George Parsons, all contributed to the shaping of the final draft. Dr. George Liik read every chapter with scrupulous care and gave invaluable advice throughout. Finally, I thank the Social Sciences Research Council of Canada for its aid in the publication of this work.

Macquarie University   George Raudzens
January 1979

Abbreviations

C.O.   Colonial Office papers at the Public Record Office, London.

C.R.E.   Commanding Royal Engineer.

G.D.   Government Document Series at the Scottish Record Office, Edinburgh.

I.G.F.   Inspector General of Fortifications.

P.A.C.   Public Archives of Canada, Ottawa.

P.P.   British Parliamentary Papers.

P.R.O.   Public Record Office, London.

R.E.   Royal Engineers.

R.G.   Record Groups at the Public Archives of Canada, Ottawa.

R.O.   Respective Officers of the Ordnance, acting as boards in charge of all departmental activities in the several military districts of the empire.

W.O.   War Office papers at the Public Record Office, London.

Chapter I

Introduction

In 1662 one hundred French regulars landed at Quebec, the first contingent of European soldiers who for the next two centuries shaped Canada's history. These garrisons, first French and then British, not only determined the fate of empires and the lines of international borders but also influenced the life and growth of the civilian populations they protected. Their story in wartime has been told, but their contributions to pioneer societies in peacetime remain too slightly documented. Thus it is well known that after 1815 the British armed forces helped colonials by spending gold and silver, providing markets for food, stores and fodder, and by building canals; army personnel in garrison towns stimulated the rise of professional classes and cultured society, and military settlers influenced the pattern and type of land settlement. They did other things as well, sometimes inadvertently—simply because they were there— and at other times deliberately. Approximately what they did has been recorded. How they did these things, and why, has only been hinted at; the same holds for their total influence on colonial development. For example, up to 1855 one little-noted military organization, the British Ordnance Department, was deeply involved with matters affecting civilian life. The activities of this body in the two Canadas between 1815 provide some of the answers to questions about the interaction of soldiers and civilians.

A study of the Ordnance Department's experience illustrates a recurring theme in Canadian history as well as the history of other lands. Beginning with antiquity, the soldiers of Rome left a physical and cultural imprint on Europe, the Near East and North Africa—up to the Rhine and above the Danube, from Syria to Britain. Between the reign of Caesar Augustus and the beginning of the third century A.D. military colonies served as the most vigorous instrument for the expansion and consolidation of the Roman way of life. To Rome these *coloniae* "had been the sinews of her civilizing power, helping her to maintain law, order and the Pax Romana."[1] Under the Former Han

---

1 E. T. Salmon, *Roman Colonization Under the Republic* (London, 1969), p. 157. For a summary of the nature, aims and achievements of Roman military colonization see Salmon, pp. 145-57. See also G. R. Watson, *The Roman Soldier* (Bristol, 1969), pp. 143-54.

1

dynasty (206 B.C. to 25 A.D.), the Chinese too had used similar techniques to expand their imperial boundaries. They first established military settlements on the northwestern frontiers; until the T'ang period of the eighth century A.D. and later, they continued to employ this form of colonization along their borders in order to expand their cultural domination in the Far East.[2] Medieval Europe also experienced military settlements. The whole edifice of post-Carolingian Christendom rested on a military basis which was central to all types of "feudalism," and on the marches—the frontiers—the role of the warrior caste was especially pronounced. From the eleventh to the fourteenth centuries the German expansion to the east and the Anglo-Norman advance into Ireland—although sustained by the general growth of peasant population—were both initially of a military character. The knight first built his castle, and around the castle grew the communities from which modern Europe evolved.[3] And in the modern era governments continued to use military means for civilian ends. It was by soldier colonies that the tsarinas Anne, Elizabeth and Catherine II strengthened the Russian grip on the Black Sea frontiers during the eighteenth century.[4] Even Americans, bound by an avowed faith in private enterprise and individualism, engaged their soldiers as pioneers in westward expansion to the Pacific. In similar ways Britain as well used military instruments to develop her overseas possessions.

There are other examples of such civilian uses of military establishments, but they have not attracted much scholarly attention as a continuing phenomenon. Admittedly, the dividing line between purely military operations—whether the conquest of new lands or the defence of existing possessions—and the consolidation of territorial control through the fostering of a distinct way of life among civilians is often blurred. Nevertheless military personnel and organizations have been engaged in helping to "Romanize" the ancient world, "Americanize" the western half of the present-day United States, and "Anglicize" portions of Canada. A treatment of this whole theme must await further investigation by historians working in many fields. For the present, it is only possible to present a case study of the British Ordnance Department in Canada, placed in the wider context of the theme by no more than brief reference to other closely related examples.

This department performed its peacetime contributions at the same time as branches of the United States army were opening the American West, as a sequel to the works of French garrisons before 1760, and within the sphere of activity of all the British armed forces. The pioneering efforts of United States army engineers on the frontier are

---

2 Denis Twitchett, "Lands Under State Cultivation Under the T'ang," *Journal of the Economic and Social History of the Orient*, II (1959), 162-203.

3 A. J. Otway-Ruthwen, *A History of Medieval Ireland* (London, 1968), pp. 109-25, and G. Barraclough, *The Origins of Modern Germany* (Oxford, 1966), pp. 258-81.

4 George Vernadsky, *A History of Russia* (New Haven, 1969), p. 124. A more detailed example of Russian military colonization under Alexander I has been presented by Richard E. Pipes in his article, "The Russian Military Colonies, 1810-1831," *The Journal of Modern History*, XXII (September 1950), 205-19.

particularly impressive and well-recorded; they serve as a gauge for the comparative achievements of the Ordnance.

Between its establishment in 1802 and the American Civil War, the West Point military academy was almost the only institution in the United States producing formally trained engineers. All officers received an education of a high standard.[5] Both during their active service and as civilians these men rendered invaluable service to their expanding, industrializing nation. As officers they were the executors of deliberate federal government policy in moving the frontiers of settlement westward. Up to the Civil War army posts and forts were usually ahead of the settlers; from these advanced bases officers and men of infantry, dragoons and cavalry worked to minimize friction between the advancing white man and the Indian, to ensure as far as possible that the nation grew peacefully and in an orderly manner. Around some of their forts grew communities which have become the towns and cities of today. During the same period the engineers improved communications by clearing western waterways of obstacles to navigation, improving harbours, marking safe wagon routes, and helping to survey railway lines.[6] Most notable were the works of the Corps of Topographical Engineers, who acted as explorers and cartographers of the far west in the 1840s. Led on three out of four major expeditions by Lieutenant John Charles Frémont, they showed the American flag in regions which their government expected to acquire but did not yet clearly possess, and mapped a vast region known previously only to illiterate Indians, fur traders and mountain men. The information the Corps supplied to the public and the trails these engineers marked out aided the flow of pioneers across the Great Plains into Oregon and California.[7]

Military influence of this and other kinds had also affected the history of North America well before the birth of the United States. From the 1660s to 1760 soldiers were perhaps only second to fur traders in shaping the development of New France. To the French crown—particularly in the eighteenth century—North American possessions were military counters to the competition of Spain and Britain. As a result New France was as much a garrison as a settlement, containing a high proportion of regular and militia fighting men. From the 1790s the beaver trade too came under increasing control of soldiers commanding the posts along the Great Lakes-Mississippi fur routes.

5   William H. Goetzmann, *Army Exploration in the American West, 1903-1863* (New Haven, 1959), pp. 12-14.
6   Francis Paul Prucha, *The Sword of the Republic. The United States Army on the Frontier, 1783-1846* (New York, 1969), pp. 330-33 and 391-95. See also Henry Putney Beers, *The Western Military Frontier 1815-1846* (Philadelphia, 1935); Ruth A. Gallaher, "The Military Indian Frontier, 1830-1835," *Iowa Journal of History and Politics*, XV, 3 (1917), 393-428; W. Turrentine Jackson, *Wagon Roads West: A Study of Federal Road Surveys and Construction in the Trans-Mississippi West, 1846-1869* (Berkeley and Los Angeles, 1952); Francis Paul Prucha, *Broadax and Bayonet: The Role of the United States Army in the Development of the North-West, 1815-1860* (Madison, Wisc., 1953); Edgar Bruce Wesley, *Guarding the Frontier: A Study of Frontier Defense from 1815 to 1825* (Minneapolis, 1935).
7   Goetzmann, *Army Exploration*, pp. 3-21 and 427-34.

In these circumstances the whole fabric of life in French North America was shaped in a military pattern. The economy was largely sustained by army expenditures, the settlers who slowly swelled the population were often disbanded soldiers, the lower social order was organized for warfare, and the colonial elite was essentially military in aspirations and character. Even the political structure was patterned on a military framework.[8] In some ways, therefore, New France was one of the best examples of military colonization in modern times.

With the conquest part of the militarized French elite departed but was replaced by a British military régime, one which continued to bolster the colonial economy through large army expenditures. Furthermore, under Colonel Guy Carleton the British even sought to maintain the old French Canadian militia structure as a bulwark against expected incursions by rebellious subjects of the Thirteen Colonies.[9] Thus, at least to the end of the American Revolution, the influence of Britain's soldiers on the Canadians was similar to that of the French garrisons.

Such influence was also exerted on colonial life by British armed forces in other places and at other times. In the 1750s the army constructed Forbes' and Braddock's roads, providing American pioneers with improved routes across the Appalachian barrier. On the other side of the globe military and naval officers organized the initial convict settlements of Australia; men such as Lieutenant Matthew Flinders and Surgeon George Bass, both of the Royal Navy, helped to explore and chart the coasts of Australia, Van Diemen's Land and Norfolk Island. During the 1840s and 50s Sir George Grey speeded the settlement of New Zealand by building a series of military roads.[10] In North America, again, the army did much to locate the United Empire Loyalists in present-day Ontario; indeed the first Loyalists were themselves soldiers and took up their Upper Canadian lands in blocks designated according to regiment. From the time of their coming to the 1860s substantial garrisons of regular troops were stationed in Upper Canada; military demand for beef, flour, fodder and timber thus stimulated pioneer agriculture and industry. Army reliance on civilian transportation and merchandising facilities accelerated the growth of provincial shipping and commerce; the medical and spiritual needs of garrison troops attracted and subsidized doctors and clergymen, thereby hastening the growth of a professional class. Sometimes the soldiers provided direct support for the construction and maintenance of Anglican churches, and in the absence of civilian peace-keeping institutions they supplied Canadians with a police force.[11] Typical of colonial expressions of thanks for such services was that of the mayor

---

8   W. J. Eccles, "The Social, Economic and Political Significance of the Military Establishment in New France," _Canadian Historical Review_, LII, 1 (Mar. 1971), 1-22.

9   Gustave Lanctot, _Canada and the American Revolution, 1774-1783_ (Toronto, 1967), p. 15 and elsewhere.

10   James Rutherford, _Sir George Grey, K.C.B., 1812-1898: A Study in Colonial Government_ (London, 1961), p. 221. See also entries for Matthew Flinders and George Bass in Douglas Pike, ed., _Australian Dictionary of Biography_ (3 vols., Melbourne, 1966), I.

11   John Philip, "The Economic and Social Effects of the British Garrisons on the Development of Western Upper Canada," _Ontario Historical Society, Papers and Records_, XLI, 1 (1949), 37-48.

of Toronto, George Gurnett, to the 15th Regiment of Foot in 1837; the City Council, he stated, would have liked the soldiers to stay, and was most grateful "for their cordial co-operation at all times with the civil authorities, particularly at the various destructive fires which have occurred in the city, at which their services in the preservation of property have been highly valuable and efficient."[12] In the area of exploration and cartography, Captain William Fitzwilliam Owen and Lieutenant Henry Wolsey Bayfield of the navy compiled the first detailed, accurate charts of Canada's inland waterways during the years after 1815.[13] Finally, many British soldiers took up permanent residence in the province after completing active service. Settling either as individuals or members of government-sponsored military colonization projects, they affected the rate and type of population growth not only in Canada but also elsewhere in the empire.[14]

The Ordnance operated within this immediate context. After 1815 it was the branch of the British military establishment which contributed most to projects of civilian utility throughout the imperial domain. Officially it was responsible for the bulk of the military property and construction work in Great Britain and all the colonies except India. Its technical experts were the Royal Engineers, Sappers and Miners, and Artillery; like the West Point trained engineers of the U.S. army, they were the best educated, most skilled men in the armed forces, some ranking among the leading engineers and scientists of the nation. By the mid-1820s this department had a near monopoly of powers to execute schemes and policies best calculated to influence civilian development in Canada. Ordnance personnel explored and carried out land and construction surveys, gave technical assistance to the young University of Toronto, provided land grants to charitable institutions and churches, and left properties which are now city parks and tourist attractions.

The department's outstanding contribution, however, was in communications improvements. Ordnance officers designed and executed the Rideau Canal project—the most expensive public work carried out in the colonies during the nineteenth century by means of military funds exclusively. Combined with the military canals on the Ottawa River, this waterway had a most important impact on provincial development. Its construction—at a cost of over a million pounds sterling—brought much money into Canada and so created jobs and stimulated agriculture, industry and commerce. Above

---

12 *The Patriot*, May 12, 1837.
13 P. G. Cornell, "William Fitzwilliam Owen, Naval Surveyor," *Collections of the Nova Scotia Historical Society*, XXXII (1959), 161-82.
14 See Helen I. Cowan, *British Emigration to North America, 1783-1837* (Toronto, 1929), pp. 65-95; Robert England, "Disbanded and Discharged Soldiers in Canada Prior to 1914," *C.H.R.*, XXVII (1946), 1-18; Norman Macdonald, *Canada, 1783-1841. Immigration and Settlement. The Administration of the Imperial Land Regulations* (Toronto, 1939), pp. 39-67; J. S. Martell, "Military Settlements in Nova Scotia After the War of 1812," *C.N.S.H.S.*, XXIV (1938), 75-105; George F. Playter, "An Account of Three Military Settlements in Eastern Ontario—Perth, Lanark and Richmond, 1815-20," *O.H.S.P.R.*, XX (1923), 98-104; George K. Raudzens, "A Successful Military Settlement: Earl Grey's Enrolled Pensioners of 1846 in Canada," *C.H.R.*, LII, 4 (Dec. 1971), 389-403.

all, in the 1830s and 1840s it served as the first improved navigation from the ocean to the Great Lakes, as the precursor of the St. Lawrence seaway.[15] The Ottawa-Rideau canal system was essential to the maintenance of the east-west Laurentian pattern of trade on which Canada's identity would grow; it helped to counter economic and technological forces threatening to draw the upper province totally into the orbit of the United States.

For Canada's economic foundations, on which a political structure distinct from that of the United States came to rest, were anchored to the St. Lawrence-Great Lakes water route; from the beginning of the French empire staple products flowed out of the continental interior along this route. By the first decades of the nineteenth century Canadian merchants, led by Montreal businessmen, were most anxious to improve this natural waterway. They wanted to draw the staple produce of the burgeoning American middle west to Montreal and at the same time counter competitive American communications projects such as the Erie Canal of New York State.[16] The builder of the Rideau Canal, Colonel John By of the Royal Engineers, saw his work as the most vital portion of a Canadian waterway which would ensure that Montreal, not New York, or some other American city, would become the leading economic metropolis of the continent. His vision proved too optimistic, but his canal did help to keep Canada's staple trade flowing in the face of American competition until the provincial community completed its own more practical St. Lawrence navigation after 1848.

As well, the Ottawa-Rideau waterway speeded the growth of settlement in present-day eastern Ontario. Agriculture, industry and urbanization all expanded rapidly from the commencement of military canal building in the 1820s. At the mouth of the Rideau Canal rose the present capital of the nation, Ottawa; it began, appropriately, as Colonel By's town, Bytown, and the canal on which it rested influenced its choice as the centre of political power. All this produced yet another effect, the John By myth which has become part of Canadian national sentiment.

When listed thus in isolation these Ordnance achievements seem impressive, but of course they constitute only a small contribution to Canada's growth after 1815. Constructive contributions to the development of frontier areas by military establishments as a whole, especially in the United States and the British Empire, have only been supplementary to the work of civilians. Yet it is precisely this which arouses interest. Governments tended to divert their armed forces into constructive activities regarded as essential when civilians were unable or unwilling to do them. Exactly how and why this happened in the past remains too often unanswered. The record of the Ordnance

---

15 Robert Leggett, *Rideau Waterway* (Toronto, 1967), p. 4. See also Robert Brown Sneyd, "The Role of the Rideau Waterway, 1826-1856" (M.A. Thesis, University of Toronto, 1965). Sneyd's work is the only detailed scholarly analysis of the impact of the Rideau Canal on Canada's development. Reference to his thesis will be made in following chapters.

16 Many historians have commented on the importance of the St. Lawrence-Great Lakes route as the communication around which Canada was formed. In particular, see Donald Creighton's *The Commercial Empire of the St. Lawrence, 1760-1850* (Toronto, rev. ed., 1956), which is perhaps the most emphatic statement of the Laurentian thesis.

provides some of the answers, answers which shed new light on generalizations previously asserted or blank spaces in scholarly writings concerning the role of the military in post-1815 Canada.

By comparison with branches of the United States army the Ordnance did not make as much of a contribution to Canada's development as its American counterparts did to that of their nation. The department was not nearly as spectacular in opening new areas for expansion. But then there were great differences between the Canadian and American frontiers. Without attempting to untangle the controverted definitions of "frontier" emanating from Frederick Jackson Turner and his followers, it can, however, be argued that the American engineers worked in vast empty regions, some claimed by foreign powers, into which pioneers were just starting to move or expected to move. The Ordnance, on the other hand, operated in British provinces with fixed boundaries and in which settlers with more or less secure ties to Britain were firmly established. These provinces, particularly Upper Canada after 1815, were frontiers insofar as they were economically and socially "under-developed" and thinly populated. But for the Ordnance the major frontier problem was related more closely to national and imperial borders than to Turnerian lines of settlement. While Americans were driving westward across the continent in the first half of the nineteenth century, Canada's chief frontier concern was the defence of the southern borders, the need to ensure that the expanding Americans did not breach the international frontier line. The imperial government lacked any ambition to expand its North American possessions; it only sought to preserve these provinces until they had strength enough to withstand American pressure unaided. Thus the principal task of the Ordnance was to safeguard Canada's international frontier; to aid the extension of settlement westward would only have added to the department's defence problems.

This defensive task had to be fulfilled during the years of the "defended border."[17] From the conclusion of the 1812 war to the Treaty of Washington in 1871 Britain and the United States clashed repeatedly over international issues, often almost to the point of armed conflict. Some disputes concerned Britain's North American territories while others were direct Anglo-American confrontations, but all threatened war in Canada. In the event of hostilities, the Americans were undoubtedly going to attempt the conquest of British North America. Despite growing resentment over colonial defence costs, the imperial government felt constrained by duty to loyal subjects to prevent such invasion. Yet to do this successfully was perhaps impossible.

The province of Upper Canada was the only major British colony beyond the protective reach of the Royal Navy. The war of 1812 had shown both antagonists that while Britain held the initiative along the American Atlantic seaboard through naval superiority, the Americans in turn held the initiative along the Canadian border. They threw away their advantage in the interior and failed to take Canada; but would they fail

---

17   C. P. Stacey, *The Undefended Border: The Myth and the Reality* (Ottawa, 1962). Canadian Historical Association pamphlet No. 1.

again in the next war? The leading British soldiers knew their American counterparts had noted their previous mistakes and did not intend to repeat them. Therefore in 1818 and 1819 the Dukes of Richmond and Wellington set out a Canadian defence plan which might enable British forces to hold the vulnerable upper province until the Royal Navy could batter the Americans into submission from the sea. This design called for a sheltered military waterway between Montreal and the Great Lakes protected by fortresses of great strength. Improved communications were essential; Canada's garrisons relied on the sealanes for all their needs, and only from the sea could they draw the power necessary to withstand the more numerous and better supplied Americans.[18]

The Ordnance had the duty to construct these vital defences, a most difficult task. Parliament did not wish to grant the huge sums the department needed, and in the end all the projects regarded by Wellington as indispensable could not be completed. Working with inadequate staff and insufficient resources, therefore, the Ordnance was not even able to carry out its military tasks, let alone consider substantial contributions to civilian improvements. Thus the total effect it exerted on Canada was limited by conditions beyond its control; there were the problems of Canadian geography coupled with the strategic concept of defence needs, lack of parliamentary support, and also defects which were inherent in the department and the whole imperial government. The changes in imperial organization after 1815 added to or accentuated such difficulties.

During the latter decades of the eighteenth century leading British statesmen had begun to campaign for increased efficiency in imperial government. They wanted to rationalize and centralize the imperial administration so it could better respond to the growing social and economic complexity of the mother-country and its posessions. For example, the creation of the office of Secretary of State for War and the Colonies in 1801 was a step in this direction. Central to the reform movement after 1815 was an effort to concentrate all executive authority in the hands of men directly responsible to Parliament. During the wars of the French Revolution and Napoleon, reforms were also made in the armed forces,[19] but after 1815 the Duke of Wellington and his circle of peninsular veterans moved against the centralization effort, resisting attempts to place the various military departments under close parliamentary control. Till his death in 1852 Wellington adhered to the conviction that military matters were for military experts only and not to be exposed to civilian, parliamentary interference. Such matters came under the exclusive authority of the royal prerogative; that is, they had to remain in the hands of the generals.

Yet simultaneously Wellington urged reforms within the Horse Guards, the Commissariat, the Quartermaster General's Department, and especially the Ordnance;

---

18   The British effort to provide Canada with effective defences after the War of 1812 is best described in two works, Kenneth Bourne's _Britain and the Balance of Power in North America, 1815-1908_ (London, 1967), especially pp. 33-52, and J. Mackay Hitsman's _Safeguarding Canada, 1763-1871_ (Toronto, 1968), especially pp. 110-29.

19   Richard Glover, _Peninsular Preparation: The Reform of the British Army, 1795-1809_ (Cambridge, 1963).

he was largely instrumental in expanding the latter department's responsibilities and powers in such a way that its organization and authority were extended to include the colonies for the first time in the 1820s. But the Duke was against proposals to place the Ordnance and other military branches under the centralized control of the War Office, which was directed by a civilian member of parliament entitled the secretary at war. Thus the Ordnance, while accountable to senior departments like Treasury, Colonial Office and Home Office, retained a wide measure of independence until 1855. The independent authority of the department grew in the colonies during a period when the Colonial Office, increasingly aware that the colonists were starting to move towards independence, began to bend to the wishes of Britain's overseas subjects and to grant them wider powers of self-government. From the mid-1820s the Ordnance had neither direct official links with Colonial Office representatives in Canada nor with the Canadians. Ordnance officers sought to serve colonial interests—in particular by providing adequate defences—without asking or thinking to ask what the colonists themselves wanted. Sometimes these officers tried to promote what they considered were colonial interests in direct opposition to Canadian desires. In short, the Ordnance was an anomaly in a changing empire, paternal in attitude while out of touch with those it wished to serve. Consequently it often saw its plans frustrated by pressure from colonists, the Colonial Office, and the imperial government.

While the Ordnance had its own internal weaknesses it also suffered from defects still pervading the whole imperial government. Britain's control over colonies was defective insofar as it was often impossible to restrain its agents on the fringes of empire. The imperial officials could commit their government to unwanted expansion, expenditure and increased responsibilities.[20] They acted on their own initiative, in response either to their private views, their better knowledge of local conditions, or both. Because communications with the outlying regions of the empire were inadequate the London authorities not only had to give such men more independent authority than was felt desirable but often the government also was compelled to accept their arrangements even when these contradicted decided policies. Thus the empire sometimes expanded against the wishes of its official decision makers and the imperial burden grew heavier and more expensive.

Officials of this type most often noted tended to be of high rank and produced major changes in the direction of imperial policy, but they also appeared among the lesser ranks of imperial agents. It was one such minor proconsul, Colonel John By, who

---

20 Outstanding examples of such "men on the spot" were Sir Stamford Raffles and Sir Andrew Clarke in Malaya, and Sir Benjamin d'Urban and Sir George Grey in South Africa. For the original explicit statement of the "man on the spot" thesis see John S. Galbraith, "The 'Turbulent Frontier' as a Factor in British Expansion," *Comparative Studies in Society and History*, II (1959-1960), 150-68. For specific examples see C. D. Cowan's *Nineteenth Century Malaya: The Origins of British Control* (London, 1961); C. Northcote Parkinson's *British Intervention in Malaya, 1867-1877* (Singapore, 1960); James Rutherford's *Sir George Grey* and J. S. Galbraith's *Reluctant Empire: British Policy on the South African Frontier, 1834-1854* (Los Angeles, 1963). Further reference to this thesis is made in Chapter V, following.

helped frustrate the aspirations of the Ordnance of Canada. The department selected this officer in haste, gave him vague orders, and sent him to Canada in 1826 to build the Rideau Canal. Colonel By decided the project entrusted to him was on too modest a scale. In the face of categorical orders by Wellington and other superiors at the Ordnance he bound the government to a much grander canal scheme, which absorbed massive unauthorized sums, roused the ire of Treasury and Parliament, and left fellow officers with insoluble Canadian problems. While the larger proconsuls usually forced territorial expansion on their government, Colonel By saddled it with burdensome expenses.

He was able to do this partly because of defective financial controls in the various departments of state. Each department tended to use its own system of budgeting, disbursement, accounting and audit, making government finances outside Britain so confused that in cases it was impossible to discover who was responsible for spending unauthorized funds or even if such funds were in fact unauthorized. Faults like these—for which the Treasury itself was in part answerable—produced situations like the Rideau Canal case, regarded by the Lords Commissioners of His Majesty's Treasury as gross misallocation of public money. The Lords failed to fix the blame clearly on Colonel By; therefore they attacked his entire department. They also redoubled their determination to reduce expenditures in the colonies even for urgent defence projects which might lead to long term savings.

The penny-pinching Treasury, in particular, is commonly blamed for forcing narrow-minded policies which in the long run created more expense than those the Lords Commissioners sought to avoid by short term economies. But this body, hounded by a tight-fisted House of Commons, was not just blindly parsimonious. Experiences such as the Ordnance spending in Canada drove it to adopt a negative attitude to colonial military costs because authorization of specific sums for specific projects had often led to inexplicable and uncontrollable drains on the public purse. These drains were due to unsystematic and cumbersome accounting and audit procedures of departments of state operating in colonies.

Such costly defects help to explain British disenchantment with the burdens of empire after 1815. The troubles of the Ordnance reflected this disenchantment. The department's officers were crippled in their effort to guarantee Canada's security. In addition to the failings of their own organization, Parliament, Treasury and Colonial Office put repeated obstacles in their way. Yet they were not merely victims of what some have described as a general British retreat from imperial responsibilities.[21] The Ordnance itself became frustrated over the difficulties involved in fulfilling Canadian

---

21 The connection between British disenchantment with the burdens of empire after 1815 and the policies which led to the withdrawal of colonial garrisons and the granting of responsible government is examined by a number of scholars, in particular R. L. Schuyler in his "The Recall of the Legions: A Phrase in the Decentralisation of the British Empire," *American Historical Review*, XXVI (1920), 18-36, and C. P. Stacey in his *Canada and the British Army, 1846-1871; A Study in the Practice of Responsible Government* (Toronto, rev. ed. 1963).

responsibilities. It, too, became disenchanted, not because colonies cost money but because colonists, far from appreciating what it tried to do for them, seemed to obstruct it at every turn. To some extent then, Ordnance representatives were driven from the colonies not only by a tight-fisted House of Commons and Treasury but also by ungrateful colonists. Because of its isolation both from the Canadians and other branches of government, the department tried to carry on in something of a vacuum. It was insufficiently responsive to changing attitudes toward the imperial tie and toward government either in Canada or Britain.

So when Earl Grey began to withdraw the legions after 1846, the Ordnance was among the first major military branches he removed from the colonies. In 1855 institutional reformers in London liquidated it altogether. Given these circumstances, the department contributed to Canadian security and material progress almost despite itself.

Ideally this story of a frustrated Ordnance in Canada should be measured against productive peacetime work of armed forces in many other times and places. Within the British Empire itself Ordnance activities need to be examined in other colonies. And a full history of the department from its origins in the days of William the Conqueror still lacks its author. Yet a case study confined to Canada is nevertheless worthwhile. Between 1815 and 1855 the department was at the height of its influence, and Canada, aside from India, was the most important of the colonies. Here the Ordnance had its biggest tasks, problems and achievements.

In summary, the department's experience is an illustration of the relationship between military influence and the progress of Canada. How and why did this military establishment impinge on provincial life in its multiple facets? And to what ultimate effect? The answers to these questions are important, for they add to an understanding of the military effect on civilian development generally, of the working of the imperial government and armed forces, and of Canada's growth.

Chapter II

The Ordnance Department

Before turning to the Canadian work of the Ordnance it is necessary to see what place it occupied in the structure of the imperial government. What were its functions and duties between 1815 and 1855? How did this department grow so that it reached the peak of its power in the 1820s but was too inefficient to survive the assault of administrative reformers in 1855?[1]

The Ordnance was directed by both soldiers and civilians.[2] At its head was the master general of the Ordnance, usually a prominent, high ranking military officer, and sometimes—as in the case of the Dukes of Marlborough and Wellington—the leading soldier in the realm. Until 1828 he was a member of the cabinet, appointed as much for political reasons as for his military qualifications. His office was usually a temporary post; Wellington's tenure of nine years was an exception. The influence of a particular master general thus depended on his rank and prestige, the length of time he served, and his administrative ability. Besides directing the Ordnance he was the chief military adviser of the government while he stayed in cabinet. After 1828 he was reduced from the position of the cabinet's expert on grand strategy to that of technical consultant on armaments and fortifications.

---

1   No detailed scholarly history of the Ordnance has yet been published. There have been only two attempts to record the department's story as a whole, Archibald Forbes' *A History of the Army Ordnance Service* (London, 1929, and N. Skentelbery's *The Ordnance Board: An Historical Note* (War Office Library Papers, Annex A to Proc. 40, 169, Sept. 1964). The former work has too many large gaps and the latter, while printed, is only three-and-one-quarter pages in length. Both are without documentation. There are a few other books dealing with branches of the department, such as T. W. J. Connolly's *History of the Royal Sappers and Miners* (London, 1855); Oliver F. G. Hogg's *Royal Arsenal* (London, 1963); and Whitworth Porter's *History of the Corps of Royal Engineers* (London, 1889 and 1915). Only Hogg's is a scholarly work. Occasional references to the department also appear in histories of the British army such as J. W. Fortescue's *A History of the British Army* (13 vols., London, 1923), XI; and Richard Glover's excellent *Peninsular Preparation: The Reform of the British Army, 1975-1809* (Cambridge, 1963). The primary sources from which a full history must be constructed are scattered, disorganized and vast.

2   This description of the organization and functions of the Ordnance is compiled from existing printed works as well as a wide reading of departmental records cited in detail in subsequent chapters.

Next in rank at the Ordnance was the lieutenant general, who was second in command to the master general. His office, however, was abolished in 1828. The third soldier in the hierarchy was the inspector general of fortifications, commonly called the I.G.F. He was the departmental technical expert, holding his post independent of the master general, maintaining a separate office in London, and supervising the three Ordnance military corps, the Royal Engineers, Artillery, and Sappers and Miners. The I.G.F.'s office produced all plans and cost estimates for construction and repair of defence works and allocated officers of the engineers and artillery and men of the sappers to Ordnance projects in the field. The inspector general was senior engineer in the armed forces, usually retaining his post for the duration of active service; thus his office carried with it longer tenure than the master general's. General Gother Mann served as I.G.F. from 1811 until his death in 1830 and field marshal Sir John Fox Burgoyne served from 1845 to 1868.[3] In fact this post survived under the War Office after the Ordnance was abolished in 1855. Two other soldiers, the colonels commanding the Royal Engineers and Artillery, were members of the department's military directorate.

Among the civilians the four chief positions were those of surveyor general, storekeeper general, clerk of deliveries, and clerk of the Ordnance. The clerk of deliveries was eliminated in 1828 and his duties transferred to the surveyor general. Each of these officers had his own department and duties, but collectively they managed routine Ordnance business. That is, they purchased, maintained, repaired, stored, and issued proper quantities of most nonperishable supplies—from buttons to howitzers— needed by land and sea forces. They inspected quality, drew up contracts for manufacture, repair and transport, kept accounts, and recorded all routine proceedings.[4] They also acted together as a board under the chairmanship of the master general; theoretically the department was directed by this board, which was officially designated the Master General and the Honourable Board of Ordnance. All action was taken "by order" of this corporate body. In fact it was more an administrative than a decision-making institution.[5]

Of the department's civilians two men outside the board were much more involved in making decisions. These were the secretary to the Ordnance and the clerk of the Ordnance chief clerk who sat on the board. The former was a permanent undersecretary with a position similar to that of James Stephen at the Colonial Office and Sir Charles Trevelyan at the Treasury. His office was at the Tower of London where he received all important correspondence, sent out directives over his signature in behalf of the master

---

3   See entries for Gother Mann and Sir John Fox Burgoyne in Leslie Stephen and Sidney Lee, eds., *The Dictionary of National Biography* (Oxford, 1917-   ). This work is hereafter cited as *D.N.B.*
4   Forbes, *History of the Army Ordnance Service*, pp. 149, 174.
5   The minutes of the board are at the Public Record Office in London in over 2,000 folio volumes under the classification War Office 47. The entries are brief and record either the reading of reports and requisitions or the issuing of orders. Both types of entries concern only routine matters such as requests for bedsteads from the Cape of Good Hope or the despatch of bandoliers to a regiment in Ireland. There are no records of decisions about major departmental policies.

general and board, and conducted liaison between the Ordnance and other departments of state, in writing and verbally. Definitions of departmental policy and documents dealing with important decisions were prepared by him. The chief clerk to the clerk, also at the Tower, dealt with similar matters but addressed himself only to officers within the department in London, usually by means of memoranda to the secretary and master general. He supplied his colleagues with detailed precedents and advice concerning major decisions. In the 1815 to 1855 period the most influential secretary was Richard Byham, holding office from 1827 to 1850, and Seth Thomas served for the longest period as chief clerk. These two officers, with the master general and inspector general of fortifications, were therefore the principal decision-makers of the Ordnance. A master general like Wellington overshadowed his subordinates and took most important decisions himself. More typically, where masters general were less vigourous and served for shorter terms, it was the secretary, the chief clerk, and the I.G.F. who ran the Ordnance.

They directed a department with many ancillary organizations. Most important were bodies known as respective officers—called the R.O.—which were based on the "out-stations" or military districts across the empire. They were replicas of the board of Ordnance comprising senior officers of the Royal Engineers and Artillery, and the local surveyor, storekeeper, and clerk in each out-station.[6] All departmental duties within their districts came under their authority. Unlike the board of Ordnance itself, their members did not act in their individual capacities but rather as a group. Each board of respective officers answered directly to the office of Ordnance in London.

Besides the R.O. the department had its own land transport branch, medical branch, education system, factories, and experimental stations. The school system is especially notable because the Ordnance did much to initiate formal technical education in Britain. The Royal Military Academy at Woolwich, founded in 1741 to train engineer and artillery officers, was the first important military and technical school in the empire.[7] It probably produced the most thoroughly trained scientists in the British Isles during the late eighteenth and early nineteenth centuries.[8] No comparable institution existed for regular army officers until the Royal Military College was founded in 1801, and even then the new school's standard of education was inferior.[9] In 1812 the Ordnance established its second school at Chatham to train the men and noncommissioned officers of the Royal Sappers and Miners.[10] Graduates of both Ordnance institutions were much respected and in great demand in the civilian sphere as well as the

---

6   After 1832 the five members of the R.O. were reduced to three by the elimination of the surveyor and clerk. See W.O. 44/28, pp. 60-61, James Smith to the Ordnance, Feb. 3, 1836.
7   Until Edward Cardwell's army reforms beginning in 1868 only Royal Engineer and Artillery officers received commissions on the basis of educational qualifications. All other army officers gained commissions by purchase.
8   Glover, *Peninsular Preparation*, p. 189.
9   Ibid., p. 207.
10   Connolly, *History of the Royal Sappers and Miners*, pp. 188, 197.

armed forces. As in the case of the West Point trained U.S. army engineers,[11] the high educational qualifications of Ordnance personnel largely explain the department's ability to make contributions to material progress at home and in frontier areas.

Also of note were the Ordnance factories and experimental stations. At the Tower of London the department had small arms repair and assembly shops, as well as a gun carriage works and barrel foundry for naval and land service artillery. In 1759 it established a powder mill at Faversham and later another at Waltham Abbey. By 1809 these mills produced three-fifths of Britain's gunpowder. The Ordnance also used them as research laboratories. Between 1804 and 1806 William Congreve the younger perfected the first modern artillery rockets at Faversham, and simultaneously Henry Shrapnell of the Royal Artillery developed the so-called "Spherical Case Shot."[12] While doing its own research, the department also set up administrative machinery for evaluating innovations of potential military use from outside its organization. In 1805 a committee of the Woolwich faculty came into being, taking on permanent form as the Ordnance Select Committee. This body was so efficient in applying new discoveries to military uses that, like the office of the inspector general of fortifications, it survived the demise of its parent department.[13]

These ancillary branches were normally controlled by the respective officers outside of London, but the Ordnance did other things as well, for aside from its normal functions, its personnel also worked for departments of state such as the Home, Colonial and Foreign Offices.[14] Thus the Home Office called on Ordnance experts to help solve problems created by technical and social change during the 1820s and 1830s. Royal Engineers worked on railway planning and management commissions, as inspectors of military and civilian penal institutions, and as construction supervisors at state dockyards and other civil works projects.[15] In 1831 Colonel John Fox Burgoyne, R.E., was placed at the head of the newly created Irish board of works.[16] Typical of the government's reliance on Ordnance officers was the appointment in January 1850 of Lieutenant Colonel William Reid, R.E., as chairman of the royal commission for the staging of the Great Exhibition. Reid supervised every step of the planning and construction of Sir Joseph Paxton's revolutionary Crystal Palace, winning the admiration of Queen Victoria for his efficiency and thoroughness.[17]

---

11 Like Woolwich and Chatham, West Point also pioneered technical education in the United States. See Goetzmann, *Army Exploration*, p. 14.
12 Forbes, *History of the Army Ordnance Service,* pp. 107-08, and Glover, *Peninsular Preparation*, pp. 67, 69, 73.
13 Skentelbery, *The Ordnance Board*, p. 2.
14 See W. O. 44/614, J. F. Burgoyne to Sir George Murray, Dec. 9, 1845.
15 W.O. 55/1931, p. 11, "Outline of the Organization, Duties and Courses of Instruction of the Corps of Royal Engineers, in the British Service," by Colonel Frederick Smith, Dec. 31, 1849.
16 George Wrottesley, *Life and Times of Field Marshal Sir John Fox Burgoyne, Bart.* (London, 1873), pp. 389-90.
17 Yvonne ffrench, *The Great Exhibition: 1851* (London, 1951), pp. 43, 143, 275.

Perhaps the department's best known civilian contribution is the Ordnance survey of the British Isles. It began in 1783 as a joint Anglo-French effort to fix the exact relative geographic positions of the observatories of Greenwich and Paris by trigonometrical calculation. Major General William Roy, R.E., began the work, and in 1791 was succeeded by two artillery officers, a Lieutenant Colonel Williams and Captain William Mudge.[18] Under these two men the trigonometrical survey expanded into a cartographic one, resulting in the publication of the first Ordnance map, a map of the county of Kent superior to any previously printed in Britain.[19] The cartographical project continued until all of the British Isles were mapped in detail, and was later taken up by the military engineers of Canada, the Australian states and other commonwealth countries. The resulting maps, available to the general public in Britain, have been kept up to date and are still regarded as the best available for all but very specialized purposes.

The Colonial and Foreign Offices also made periodic use of Ordnance experts. For example, Royal Engineers and Sappers provided technical aid to diplomats in determining the boundaries between the United States and British North America during the 1830s and 1840s. In 1839 an Ordnance party under Colonel Mudge conducted a topographical survey of the disputed area between the state of Maine and the province of New Brunswick in an effort to improve the British bargaining position against the Americans.[20] Other groups of Ordnance men worked as technical members of the North American boundary commission between 1843 and 1846, helping to implement the terms of the Webster-Ashburton treaty (1842) which settled the Maine-New Brunswick border. The Foreign and Colonial Offices used these men because local civil engineers were too scarce and of insufficient skill for such important work.[21] Similar reasons prompted the appointment of lieutenants M. Vavasour and H. Warre of the engineers to reconnoitre the line of the 49th parallel between the Great Lakes and Rocky Mountains in 1845; their work was later completed—after the abolition of the Ordnance in 1855—when both British and American engineer officers finished the survey of the Anglo-American boundary to the Pacific.[22]

Simultaneously the American Topographical Engineers were also surveying these boundaries. In 1845, while Royal Engineers were marking out the border of New Brunswick, Topographical Engineers under Lieutenant Colonel James Kearney were surveying the Great Lakes border region and coming into occasional contact with Ordnance officers who had similar duties.[23] As well, the Topographical Engineers

18 See entry for William Mudge in *D.N.B.*

19 W.O. 44/614, Precis relating to the survey of England and Wales, by Lieutenant Colonel Thomas Colby, Jan. 7, 1834.

20 Connolly, *History of the Royal Sappers and Miners*, p. 347.

21 Ibid., p. 416, W.O. 44/506, James Stephen to R. Byham, Apr. 28, 1846, and ibid., A. W. Addington to Byham, Jan. 21, 1946.

22 David Veitch, "The Royal Engineers and British Columbia," *The Royal Engineers Journal*, LXXII (1958), 107-17.

23 W.O.44/506, James Stephen to Byham, July 9, 1845, and James Kearney to Major General Sir Richard Armstrong, May 19, 1845.

surveyed the far western boundary between the United States and British North America, and the United States-Mexican border after 1848.[24]

Sometimes the Colonial Office called on Ordnance engineers to help build new settlements by providing initial land surveys. Captain Edward Charles Frome, R.E.—who worked on the Rideau Canal between 1827 and 1833—took out a party of sappers to South Australia in 1839; his orders were to survey town sites, country properties and roads in advance of the settlers who would purchase the land, so that the colony would be settled methodically and rationally, according to the precepts of Edward Gibbon Wakefield. Frome also acted as colonial engineer supervising construction of many vital civil works. In 1844 governor Captain George Grey officially commended this officer's contributions to "scientific" colonization.[25] Other Royal Engineers carried out land surveys in Van Diemen's Land during the late 1840s and 1850s and helped Australian colonial governments with civil works.[26]

Another extra-departmental activity of Ordnance officers was the promotion of scientific research. Men of the engineers and artillery built and manned meteorological and magnetic observatories at the Cape of Good Hope, St. Helena and Toronto. Captain J. H. Lefroy, R.A., did especially valuable research at the Toronto observatory between 1840 and 1853, and then turned his scientific establishment over to the University of Toronto.[27] Sometimes Ordnance officers helped to stimulate the advancement of science in the colonies in their private capacities. In 1832 the editor of the Upper Canadian newspaper, the Kingston *Chronicle*, praised Captain Richard Bonnycastle, R.E., who, ". . . by his professional and scientific acquirements has contributed considerably during his residence in Kingston to promote the discovery of our Canadian resources, and has published many valuable documents on geological researches, which are likely to develop the natural history of the country and are highly important in the advancement of ulterior discoveries."[28] Throughout the empire other Ordnance men performed such services and so directly contributed to civilian progress.

---

24 Goetzmann, *Army Exploration in the American West*, pp. 153-208.

25 Connolly, *History of the Royal Sappers and Miners*, p. 342. See also entry for E. C. Frome in *Australian Dictionary of Biography*, I. This officer demonstrated his high technical competence by publishing one of the first and most comprehensive descriptions of the Rideau Canal, "Account of the Causes which led to the Construction of the Rideau Canal, connecting the Waters of Lake Ontario and the Ottawa; the Nature of the Communication prior to 1827; and a Description of the Works by means of which it is converted into a Steam-boat Navigation," *Papers and Subjects Connected with the Duties of the Corps of Royal Engineers* (London, 1837), pp. 73-102.

26 W.O. 6/101, pp. 92-96, H. Merivale to G. Butler, July 11, 1851.

27 W.O. 44/514, "Memorandum in reference to the maintenance of the Observatory at Toronto" by Captain J. H. Lefroy, Nov. 19, 1852. More detail about Lefroy's work at Toronto is provided in Chapter X following.

28 Kingston *Chronicle*, Apr. 28, 1832. Bonnycastle was one of the most outstanding Ordnance officers serving in Canada. As commander of Canadian militia he won a knighthood for the defence of Kingston against Americans during the rebellion crisis of 1837 and 1838. He also published five books, four about British North America, the best known of which is *The Canadas in 1841* (2 vols., London, 1842); these volumes helped to attract British immigrants to the provinces. For the most

Such were the functions of the Ordnance during its last four decades. It was a department with multifarious, even esoteric, duties. In fact its workings appear to have been more complex—if less critical for the imperial government—than those of senior departments like the Colonial Office. The latter, while altered by a series of reorganizations after 1815, still seemed to have a more easily definable range of functions than the Ordnance.[29] How, then, did the structure of the Ordnance evolve to its maximum extent after 1815, yet without the strength to survive past 1855?

The roots of the department can be traced back almost to 1066 and particularly to the establishment of the royal arsenal at the Tower of London.[30] By 1414 Nicholas Merbury became the first official Master of the King's Ordnance; it appears that subsequently, with the growing importance of guns, the men at the Tower began to develop a formal departmental organization. Queen Elizabeth recognized the value of this heavy weapons branch, reorganized it in 1597, and appointed her favourite, Robert, second earl of Essex, the Grand Master of the Ordnance.[31] Charles II carried through another reorganization in 1683 to adjust the department to expanding military needs.[32] Yet these reforms failed to keep pace with accelerating demand for military technical services; during the eighteenth century more and more critics attacked the Ordnance with accusations ranging from misconduct to corruption.

In 1780, when the whole imperial administration was under assault from reformers, Edmund Burke was especially critical of the department and forced it to improve its finances. Burke's reforms failed to eradicate the confusion of Ordnance accounts.[33] As its annual expenditures grew, so parliamentary criticism increased. At the same time its areas of responsibility became blurred, overlapping with those of other departments, and its capacity to carry out duties declined. Charles Lennox, the third duke of Richmond, worked hard to remedy these defects while master general from 1782 to 1798,[34] but less able successors during the years of war with France again left the Ordnance to struggle along in confusion. Lacking strength to cope with wartime demands, the department found it could not supply enough weapons, munitions, engineers and artillery officers. The Duke of York considered it so inadequate that in 1799 he created a rival corps of engineers, the Royal Staff Corps.[35] The Treasury, the

comprehensive sketch of his life, see J. E. R. Munro's unpublished manuscript, "Sir Richard Henry Bonnycastle, Lieutenant Colonel Royal Engineers," at the Archives of Ontario, Toronto. For further details about his Canadian contributions see Chapter X, pp. 131 and 140.

29 See J. C. Beaglehole, "The Colonial Office, 1782-1854," *Historical Studies, Australia and New Zealand*, I (1940), 170-89; Charles Jeffries, *The Colonial Empire and Its Civil Service* (Cambridge, 1938); Helen Taft Manning, *British Colonial Government After the American Revolution, 1782-1820* (New Haven, 1933); and Ralph B. Pugh, "The Colonial Office, 1801-1925," *The Cambridge History of the British Empire*, III: *The Empire Commonwealth, 1870-1919* (Cambridge, 1959).

30 Forbes, *A History of the Army Ordnance Service*, p. 6.

31 Skentelbery, *The Ordnance Board: An Historical Note*, p. 1.

32 Forbes, *A History of the Army Ordnance Service*, p. 96.

33 Ibid., pp. 170-72.

34 Glover, *Peninsular Preparation*, p. 38.

35 Forbes, *A History of the Army Ordnance Service*, p. 177; and Glover, *Peninsular Preparation*, p. 102.

Secretary of State for War and the Colonies, and the East India Company all established their own munitions supply organizations.[36] By 1815 the Ordnance was both over-strained and discredited.

With the passing of the stresses of war, however, it began to regain strength and efficiency. After 1821 the Duke of Wellington rebuilt it, expanded its functions, and placed it at the peak of its influence.

Traditionally Wellington was the arch-conservative in matters military as well as political. It has been asserted that he blocked all progress in military reform and was thus largely responsible for the Crimean War disasters.[37] Perhaps he did regard the seventeenth century Brown Bess musket as the ultimate weapon,[38] but as his defenders have long argued he was far from a total reactionary. Instead of rejecting new technical innovations he supported the adoption of Shrapnell's artillery shells, used the first portable military suspension bridges, employed prefabricated timber hospital buildings, and advocated the construction of military railways. More important, he insisted on absolute efficiency within the military branches he directed even while refusing to change their basic structure. He had the outstanding talent for making old institutions and procedures work in new circumstances. Wellington took Britain's eighteenth-century war machinery, strengthened it along traditional lines, and produced the most efficient army his nation ever placed in the field.[39]

It was this talent for rejuvenating outdated organizations which the Duke applied to the Ordnance. He took up the position of master general in 1819, probably for political reasons because it gave him a seat in the cabinet. Initially he showed no inclination to change the department. Just before assuming his new office he wrote the incumbent master general, Lord Mulgrave, asserting that "I am too well aware of the benefits which the Ordnance Dept. has derived from your Superintendence. I have myself derived too much benefit from it to think of altering anything of which the course of time may not render an alteration necessary."[40] From 1819 to 1821, therefore, Wellington apparently followed a policy of laissez-faire. But then he began a campaign of major reforms.

After 1815 one of the big weaknesses of the Ordnance was lack of power to control its own personnel, especially in the colonies, recently increased in numbers. Aside from its duty to supply artillery and munitions to the army and navy, the department was theoretically responsible for all construction and maintenance work on military properties and buildings. But not all the construction sites or buildings were under Ordnance control, particularly in the colonies. Thus engineers and artillery officers had to work for

---

36 Glover, *Peninsular Preparation*, pp. 49-56, 63.
37 Richard L. Blanco, "Reform and Wellington's Post Waterloo Army, 1815-1854," *Military Affairs*, XXIX, 3 (1965), 123-31.
38 Herbert Maxwell, *The Life of Wellington. The Restoration of the Martial Power of Great Britain* (2 vols. London, 1900), II, 136-37.
39 S.P.G. Ward, *Wellington* (London, 1963), pp. 104-05.
40 Apsley House, London, Wellington MSS: Wellington to Mulgrave, Nov. 1, 1818.

other departments instead of their own in places where the Ordnance had no boards of respective officers or physical establishments. Responsibility without power caused confusion and inefficiency. For example, in 1815 three Royal Engineers at Montreal complained they were so overworked by other departments they deserved extra pay. Their duties, they stated, consisted ". . . independent of the Planning and executing all works of Fortification, & making surveys of the Country; in erecting and repairing Boats; and in fact the whole of the Duty of the Quarter Master Gen^ls & Barrack Departments. Besides this the Gun carriage Establishment has been again thrown on the Engineers. . . ."[41] Wellington himself described the unsatisfactory position of the Ordnance abroad as follows:

> The mode in which I understand this business has been conducted hitherto is this: The Governor, or Commander in Chief in the Colony, after consulting with the Officers of Engineers, or other Persons as he may think proper, transmits to the Treasury an account of the necessity of any particular work or Building, such for instance as the Citadel of Quebec, with Plans and an Estimate of the probable Expence of constructing it; and the Treasury give their concurrence for its construction. . . .
>
> These Works are in general planned, estimated and executed by the Officers of Engineers stationed in the Colony and paid by the Ordnance Department; but they are not necessarily constructed by these Officers; and at all events they do not at present act under the Ordnance Department in any matter relating to these Works & Buildings. For instance I believe that to this moment this Department has no knowledge whatever of the Citadel of Quebec.[42]

As a first step to remedy these weaknesses, therefore, in 1821 the Duke procured an Ordnance Estate Vesting Act, by which the bulk of all imperial military properties were placed under Ordnance ownership.

Before 1821 large quantities of buildings and land for a variety of military purposes had accumulated in the imperial domain. These properties were acquired under acts of parliament, under legal conveyances directly to the monarch and his heirs, or under conveyances to individual officers of the several military departments. In the colonies governors set aside military reserves, using them at their discretion. Legal titles and management arrangements were chaotic. The 1821 Vesting Act eliminated the confusion by conveying the properties "To the Principal Officers of the Ordnance for the time being in trust for His Majesty his heirs and successors for the use and service of the Ordnance Department. . . ."[43] For the first time one department of state gained the power to defend military lands and buildings from trespass, to sell or lease properties no longer needed, and to purchase new sites for forts, barracks and other works with clear

---

41 W.O.55/860, p. 131, Captain Samuel Romilly, Captain W. Dixon and Captain J. De Gaugreben to Lieutenant General Gother Mann, June 30, 1815.

42 W. O. 44/265, pp. 322-23, Minute by Wellington, Mar. 23, 1822.

43 W.O. 44/28, pp. 38-42, Minute by James Smith, Feb. 1836.

title. There was no difficulty with the Act in the British Isles. But while it was intended to cover the whole empire, colonies with legislatures denied that it applied to them. The Ordnance struggled until the 1840s before it got full legal control over military possessions in these colonies. Nevertheless by 1821 the respective officers became important landlords in the out-stations and Ordnance power was increased.

To make maximum use of the Vesting Act, Wellington also moved to absorb the stores and supply branch of the Commissariat and the Barrack Department. Since the Commissary of Stores was supplying the armed forces with the same equipment as the Ordnance, he argued, the latter should do the job alone, thus eliminating expensive, wasteful duplication of services.[44] The Commissariat could be left in charge of perishables such as food, fodder and fuel, which the Ordnance did not issue.[45] As for barracks, Ordnance personnel built and repaired them, and so it was logical the Ordnance should manage them. Expressing his conviction of the need to expand his department's authority and rationalize its functions, the Duke stated that

> I cannot, however, but be of opinion that great advantage would be derived to the public interest by placing not only barracks at home and abroad, but all military works in the colonies, under the immediate superintendence and direction of the Ordnance; confining the governors abroad to a specified sum unless by the sanction of the Board, and the Ordnance to a specified sum unless by the sanction of the Treasury.[46]

This expansion, he admitted, would increase his department's expenditures. Ordnance expenditures for construction and repair would rise from the 1822 total of £53,000 sterling for the whole empire to some £68,000 sterling for Canada alone,[47] but this would only constitute a simplification of military accounts. As Wellington somewhat tritely remarked, the department which had the duty to care for all "the defensive Works of the Empire" was best equipped to do so and should have the necessary power.[48]

Some of Wellington's associates disagreed. Inspector general of fortifications Gother Mann officially protested that innovations would only cause trouble, for the old ways worked well enough.[49] The Duke's will, however, prevailed, and a Treasury minute of May 24, 1822, put the Vesting Act, in its fullest scope, into effect beginning the following year.[50]

The conditions of transfer of all military property to the Ordnance were set out in regulations in 1823, and with clarification, again in 1826. The list of transferred items

44 Arthur Wellesley, *Despatches, Correspondence, and Memoranda of Field Marshall Arthur, Duke of Wellington, K.G.* (8 vols. London, 1867-80), I, 171-72; Wellington to Arbuthnot, June 1, 1821.
45 Ibid., pp. 197-202, "Memorandum: Transfer of Barrack Department and Commissary of Stores to the Ordnance," by Wellington, Dec. 1821.
46 Ibid., p. 172, Wellington to Arbuthnot, June 1, 1821.
47 W.O. 44/265, p. 322, Minute by Wellington, Mar. 23, 1822.
48 Ibid., p. 324.
49 W.O. 44/265, pp. 328-30, Gother Mann to R. H. Crew, Apr. 8, 1822.
50 W.O. 44/175, pp. 137-43, W. Griffith to the R.O. at all outstations, Apr. 21, 1826.

included everything from water tanks to fortresses.[51] Thus from the mid-1820s the Ordnance theoretically gained control over all military establishments on which its personnel normally worked. Boards of respective officers throughout the empire now ensured that the department had all the facilities its men needed, and all departmental functions, except when officers were loaned to other departments for special tasks, were financed under one set of parliamentary estimates, or one budget. Where previously the Ordnance had been forced to scatter its manpower among other departments of state, after 1821 Wellington removed all extraneous authority from the sphere of departmental activity. In other words, he made an already independent department more autonomous still. It remained subordinate to senior departments—the Treasury, Home Office and Colonial Office—but as long as it got its annual estimates through Parliament it had the authority to carry out its duties in its own way. During the Duke's tenure as master general the Ordnance used its expanded independence to such good effect that it escaped the criticisms of administrative reformers which had been the norm prior to 1815.

The structure Wellington thus erected remained basically unaltered until 1855; nevertheless the department retained defects. Some Ordnance practices were antiquated and awkward, its internal finances remained confusing even to itself, and minor reorganizations were necessary after 1827. Most important, the Duke's successors lacked his energy and held office for short periods. Wellington resigned in December 1827, to be succeeded by Henry, Marquis of Anglesey, who in turn was followed only months later by William, Viscount Beresford. The latter was the last master general to hold a cabinet seat.[52] Towards the end of the 1820s Ordnance critics grew vocal again, and now masters general, no longer privy councillors, had neither the position nor political prestige with which Wellington was able to silence opponents.

The critics initially renewed their assault because the Ordnance spent too much. Like the generality of reformers in and out of Parliament during the first half of the nineteenth century they were particularly upset by anything which seemed wasteful or extravagant expenditure of the taxpayers' money. Large military spending for colonial defence was especially unpopular, and the Ordnance became the biggest spender in this category.[53] Wellington's reorganization, in fact, made the Ordnance the third most expensive of all the military branches and—after the army and Commissariat—the most vulnerable. Army expenditures were decided by the cabinet, and the Commissariat was a tool of the Treasury, but the Ordnance, cut off from cabinet or Treasury contact, seemed to work in mysterious and wasteful ways.

In 1848 Earl Grey wrote a critical summary of the total British outlay for colonial barracks and fortifications since 1815, stating that the Ordnance alone had spent four million pounds. He added that just prior to 1848 the department had been spending in

---

51 Ibid., pp. 138-43, List of properties transferred to the Ordnance, Jan. 1826.
52 See *D.N.B.* entries for Henry, Marquis of Anglesey, and William, Viscount Beresford.
53 See R. L. Schuyler, "The Recall of the Legions," and Stacey, *Canada and the British Army.*

excess of £300,000 per annum.[54] His figures were meant to impress colleagues who felt the Ordnance was financially extravagant, but even so he underestimated the totals. In the two Canadas alone the department's expenditures between 1825 and 1834 came to £1,548,102,[55] and to £1,220,903 between 1836 and 1843.[56] In other words, during sixteen of Grey's thirty-three years the department spent £2,769,005 in Canada alone, almost three-quarters of Grey's sum for double the period and the whole empire.

Examining these Canadian figures in detail, in 1825 the department spent £46,694 out of a total military expenditure of £237,767. For the next three years the Ordnance sum rose rapidly; in 1828 it represented £195,641 out of £387,145. The peak year was 1830, when the Ordnance spent £339,533 out of £521,094.[57] Thereafter the department's total declined to £52,035 in 1837, rose again to £276,995 in 1839, and declined again to £134,097 in 1843.[58] Furthermore, because of confused accounting, these figures fail to show the large additional expenditures which were listed in the parliamentary estimates of other departments but were actually the responsibility of the Ordnance. This was the case with the vastly expensive Rideau Canal. Due to defective planning the canal was ostensibly a Colonial Office expense, but the Ordnance was responsible for construction costs. In 1830 alone over £200,000 was spent on the canal.[59] When this is added to the official Ordnance figure of £339,533, the total exceeds all other 1830 military costs in Canada put together.

All this caused growing concern among administrative reformers, especially when it became known that total Ordnance figures did not always appear in parliamentary estimates. Evidently the department could sometimes spend public money on its own authority in pursuit of policies contrary to those of cabinet ministers directly responsible to Parliament. At a time when the reformers were tightening parliamentary control over the civilian branches of the imperial administration,[60] the Ordnance still operated

---

54   Arthur G. Doughty, ed., *The Elgin-Grey Papers, 1846-1852* (4 vols. Ottawa, 1937), I, 249, Grey to Elgin, Nov. 10, 1848.
55   C.O. 42/257, pp. 21-22, "Statement of Expenditure incurred by Great Britain from the Military Chest on account of the Military Service in the Canadas from the year 1825 to 1834, inclusively . . ." (hereafter referred to as "British military expenditure in Canada, 1825-1834").
56   P.P. 1844 (304) XXXIV, 23, "Return of the Charges incurred on account of the *Canadas*, in respect of the Army, Navy, Ordnance, and Commissariat, in each of the Years 1835 to 1843, inclusively . . ." (hereafter referred to as "British military expenditure in Canada, 1835-1843").
57   C.O. 42/257, pp. 21-22. British military expenditure in Canada, 1825-1834.
58   P.P. 1844 (304) XXXIV, 23, British military expenditure in Canada, 1835-1843. Taking three sample years, the proportional expenditures of Army, Ordnance and Commissariat were as follows:

	Army	*Ordnance*	*Commissariat*
1836	£ 74,474	£ 53,711	£ 35,058
1839	£465,231	£276,995	£ 863,494
1842	£481,189	£157,794	£199,017

59   For the annual cost of the Rideau Canal see Chapter VII following.
60   For an introduction to the long and complicated history of British administrative reform, see Llewellyn Woodward, *The Age of Reform, 1815-1870* (Oxford, 1964).

outside parliamentary authority. So too did the Horse Guards, the Quartermaster General's department, and other military branches. It was obviously time to modernize the imperial war machine and to bring the military departments into line with nineteenth century administrative practices. The reformers therefore proposed to eliminate the existing structure and replace it by a single ministry of war.

In 1834 the Duke of Richmond chaired a parliamentary commission for consolidating the military departments, and one expert witness, Colonel John Fox Burgoyne, put forward the idea of a single war ministry.[61] He advocated the abolition of the Ordnance, his own department, in the interests of unified control and efficiency. Other reformers took up the idea and it appeared in the reports of the 1835 and 1837 parliamentary commissions on military reorganization as well as in the report of the select committee on the Ordnance in 1849.[62] The secretary at war, who until 1855 was merely the parliamentary spokesman of military departments over which he had almost no influence, was most often nominated as head of the proposed new ministry.

By 1849 most British statesmen were agreed that this ministry had to be formed as soon as possible. The leading advocate was the secretary at war, Fox Maule; he singled out the Ordnance as the prime example of the ills inherent in the old system. As he stated,

> Whatever the Ordnance office may have been it is now notoriously a clog upon the Military Service. Its departments are *cumbrous*, its business is transacted in a complicated manner, its accounts are kept in an inferior way to those of other departments and the impediments which obstruct one at almost every time in transacting the commonest business within the Ordnance are proverbial in all departments that have to do with them.[63]

To get his way, however, Maule and his associates had to overcome the great obstacle Wellington.

The Duke, with peninsular veterans like Lord Anglesey, strenuously opposed the idea of a monolithic war department. In 1838 he argued that the old system contained checks and balances designed to safeguard Britain's constitutional monarchy, to preserve the royal prerogative, and to keep military decision-making out of the hands of ignorant, irresponsible parliamentarians with no experience in soldiering. The old system had brought much honour to the empire, was still working smoothly, and could be relied upon in the future.[64] Wellington especially defended the Ordnance. In 1849 he cited the report of an 1828 House of Commons select committee, which had held up the Ordnance as a model of efficiency; when the Navy was reorganized shortly after it

---

61 Wrottesley, *Life and Times of Field Marshal Sir John Fox Burgoyne*, pp. 397-98.

62 G.D. 45/8/66. "Memorandum on the Report of the Select Committee on the Ordnance Department of 1849" by H. Hardinge, Jan. 12, 1850.

63 Ibid., Confidential remarks on Wellington's report on the report of the select committee on the Ordnance of 1849, by Fox Maule, Jan. 1850.

64 Ibid. Confidential memorandum to Viscount Melbourne on proposed reforms in the organization of the military departments, by Wellington, Jan. 4, 1838.

had adopted reforms from the Ordnance. Since then the department had expanded its activities while reducing its staff, and was still rendering completely satisfactory service in all respects.[65]

Wellington's opposition was too formidable for the reformers. Fox Maule admitted as much when he wrote that no fundamental military changes could be made while the Duke lived. Writing of Wellington and Anglesey, he expressed reluctance to initiate even the most urgently needed reforms which might "disturb the setting sun of either gallant Soldiers [sic]."[66]

But the reformers were not completely immobilized by Wellington's resistance. They were at least able to make a start. In 1846 Earl Grey set out to reduce the colonial garrisons and in the process to dismantle Ordnance branches abroad. In 1849 he arranged with the legislature of New South Wales the transfer of most Ordnance property to the colony, and followed up with similar arrangements elsewhere. By 1855 the department's overseas domain was reduced almost to the insignificance of 1821. Yet because of Wellington, the Ordnance managed to survive for three years after its protector's death in 1852.

Between 1815 and 1855, then, the Ordnance was independent, complex, and diverse in function. It controlled most imperial military property, supplied the armed forces with heavy armaments, nonperishable equipment, fortresses, and other buildings, and loaned its technicians and scientists to civilian departments of state for the promotion of useful public works. It was the best-equipped military department to further material progress in the empire. At the same time, while Wellington's reforms increased its capacity to aid civilians, its resultant enhanced independence attracted the criticisms of centralizing reformers. Thus the position of the Ordnance became increasingly isolated and the department found it increasingly difficult to carry out its duties. After 1846 it began to crumble under reformer pressure. Its rise and decline, its potential and limitations as a military contributor to civilian advancement, are shown clearly by its Canadian experience.

---

65 Ibid. Confidential memorandum on military reorganization, by Wellington, Nov. 30, 1849.

66 Ibid. Confidential remarks on Wellington's report on the report of the select committee on the Ordnance of 1849, by Fox Maule, Jan. 1850.

Chapter III

The Canadian Setting

The part the Ordnance played in Canada was determined by the colonial environment as well as by the department's changing position within the imperial government. Canada exerted four general effects. First, because of geography and slow economic and social expansion—especially in Upper Canada—the colonials were virtually defenceless against American invasion without massive British support; hence the military burdens of the Ordnance were so heavy it had little resources to spare for civilian aid. Second, because Canada—as a political entity distinct from the United States—was grounded on east-west communications, the greatest Ordnance task was communications improvement. Third, because canals were the most efficient means of transportation in North America until the railway era began in the 1840s, canal construction and maintenance was the department's chief concern. Fourth, because Canadians were gaining self-government faster than any other British colony, they were able to frustrate the Ordnance whenever its interests clashed with theirs. At times the colonials blocked even those Ordnance efforts aimed at improving their own military security.

By 1815 Canada was Britain's most important colony excepting India, which had a place outside the ordinary imperial organization. The Canadian region, an upper and lower province until 1841 and a united one thereafter, contained the largest population, had the richest economy, and provided the biggest problems for the imperial government. Thus the Ordnance was more deeply involved here than in most other colonies. By contrast its commitments in the adjacent Atlantic provinces of Nova Scotia, New Brunswick, Prince Edward Island and Newfoundland, were much smaller. The department was involved in only two major tasks in the maritimes, the construction and upkeep of fortifications in and around the great Halifax naval base and an effort to build a military road overland from Halifax to Quebec. Even so the Halifax fortifications were already established by the time the Ordnance set up its own branches in British North America. After the decision to enlarge the central fortress in 1826, steady construction went on between 1828 and 1861, ultimately costing the government £233,882. The nature of this work was routine; it involved no serious financial or technical difficulties.[1]

---

1 See Harry Piers, *The Evolution of the Halifax Fortress, 1749-1828* (Halifax, 1947), publication of the

26

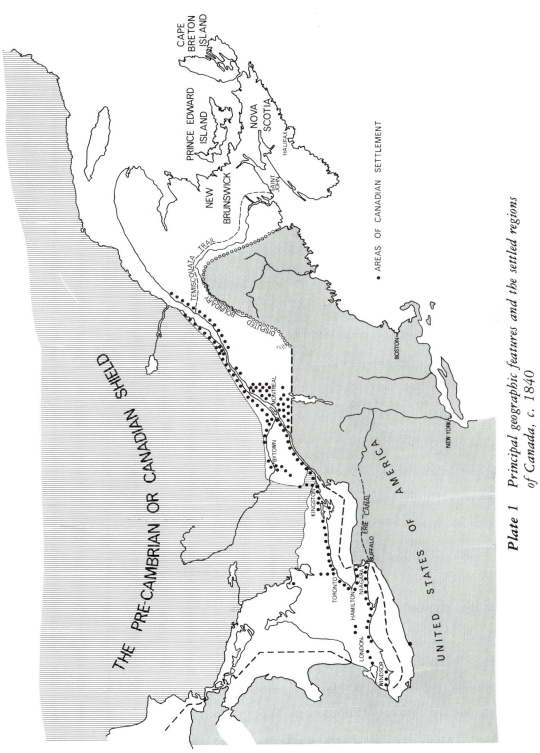

Plate 1 *Principal geographic features and the settled regions of Canada, c. 1840*

The Halifax to Quebec road, however, was a more critical project because it was needed to improve Canadian military security.

For some five months each year the frozen St. Lawrence River blocked all access to the Canadas by water. From the end of the American Revolution the British had recognized the need for an alternative land route to the inland provinces over which military aid could be rushed in the event of an invasion by the Americans in winter. The Temiscouata road, a rough trail running from the mouth of the Saint John River through the New Brunswick wilderness to the St. Lawrence near Quebec, had come into use before 1783 as an emergency mail route; in the first half of the nineteenth century this trail was the sole alternative to the St. Lawrence as a road to Canada. During the War of 1812, in the winter of 1813 and 1814, army and navy reinforcements used it, getting through to Quebec with much difficulty. After 1815, with some slight improvements, it became a more-or-less permanent winter Post Office communication. When the Canadian rebellions broke out in 1837 and touched off an Anglo-American crisis, British troops again used the Temiscouata road, now sufficiently improved to pass horse-drawn sleighs.[2]

But while the route was valuable in military emergencies, it had two major defects. Until the Webster-Ashburton treaty of 1842 it ran through disputed territory, and thereafter, despite the fact that British statesmen managed to win a boundary which placed it within their territory, the trail was still too close to the border for safety in wartime; it was also much too rough before and after the treaty. Thus a properly constructed heavy vehicles road with fortified defence points was necessary. It was up to the Ordnance to construct such a road.

In 1825 a commission of Ordnance officers under the chairmanship of Sir James Carmichael Smyth recommended the commencement of the Halifax to Quebec military road project while surveying British North American defence needs. The commission members stipulated that this road had to be as far from territory claimed by the Americans as possible and that it had to be adequately protected by fortifications.[3] Scarcity of resources, however, produced long delays. In 1831 the master general, Sir James Kempt, urged that the project should be started as soon as possible,[4] but preliminary surveys and planning did not begin until 1844. In 1845, while Ordnance offices were still marking out possible routes, the London authorities, responding to pressure from colonists, altered the project and the Royal Engineers began surveys for an intercolonial railway instead.[5] This new scheme was to be jointly financed by Britain

Public Archives of Nova Scotia; Thomas H. Raddall, *Halifax, Warden of the North* (London, rev. ed., 1950), pp. 167-68; Bourne, *Britain and the Balance of Power*, pp. 47-48; C. P. Stacey, "Halifax as an International Strategic Factor," *Canadian Historical Association Report* (1949), 46-55.

2   C. P. Stacey, "The Backbone of Canada," *C.H.A. Report* (1953), 1-13.

3   C.O. 42/208, Report of a commission of Royal Engineer officers under the chairmanship of Colonel Sir James Carmichael Smyth on the defences of British North America, Sept. 9, 1825 (hereafter cited as Smyth commission report).

4   R.G. 7, G.1, Vol. 22, pt. 1, Sir James Kempt to Viscount Goderich, Jan. 29, 1831.

5   The work of the Ordnance on the Halifax to Quebec road and intercolonial railway has not yet been

and the North American provinces, but disagreements over routes produced a stalemate by 1852; the railway plan was not renewed until the next decade. The Halifax to Quebec road failed to get under way before 1855.

Canadian security problems dominated Ordnance activity in British North America. In the first instance geography was responsible for this. Canada was unique among British possessions because all of it was landlocked in winter and its western half, beyond Montreal, was landlocked all year round. British merchant and warships, the indispensable carriers of imperial power, could ascend the St. Lawrence only as far as Quebec, or at best Montreal. Only small river craft could pass beyond the rapids and obstructions until the opening of the Ottawa-Rideau waterway in 1834 and the first St. Lawrence navigation in 1848. Even after 1848 just small sea-going ships, with a draught of less than nine feet, could reach Lake Ontario in summer.[6] In addition, past Montreal the St. Lawrence route ran along the American border and could be cut entirely during an Anglo-American war. To this extent Canada was isolated from British commercial and military power and, conversely, at the mercy of the expanding Americans.

This situation existed because Canada consisted of a chain of settlements in river valleys and along lake fronts, most of them adjacent to the United States. During the seventeenth century the French fur traders built their habitations on the banks of the St. Lawrence which, like the Mississippi, Hudson Bay, and less so the Hudson River, was one of the main natural water entries into the interior of the continent. To fur traders, in fact, the St. Lawrence was initially the best of these waterways. Skilled men in canoes or small craft, like bateaux—capable of getting past rapids—could travel up the river to the Great Lakes either by going directly to Lake Ontario or, by using the Ottawa and Mattawa Rivers, Lake Nipissing and the French River, to Georgian Bay on Lake Huron. From the Lakes, river systems led south to the Mississippi Valley, west to the Red and Saskatchewan River valleys, and north to Hudson Bay. The French followed all these routes, but especially those to the west and north because they led through North America's best fur areas, the region of the Precambrian or Canadian Shield.[7] This Shield of very ancient rock and countless lakes is heavily wooded with conifers below its subarctic latitudes, rich in fur-bearing animals and minerals, but unsuited to agriculture. Extending in a broad belt around Hudson Bay it covers the maritime provinces and most of present-day Quebec and Ontario. East of Lake Huron, which marked the western boundary of pre-confederation Canada, the Shield confines arable land to the shores of the St. Lawrence and Richelieu Rivers, to the so-called Eastern Townships

---

documented by historians. While the relevant documents are too numerous to cite in detail most of them are found in W.O.1/539, W.O.1/540, W.O.1/541, W.O.1/542, W.O.44/39 and W.O.44/46.

6   See G. P. de T. Glazebrook, *A History of Transportation in Canada* (2 vols., Toronto, rev. ed., 1964), I, 59-96.

7   See Harold A. Innis, *The Fur Trade in Canada: An Introduction to Canadian Economic History* (Toronto, rev. ed., 1967), pp. 383-402.

below the St. Lawrence to the east of Montreal, to the Ottawa Valley, to the belt along the north shore of Lake Ontario roughly south of a line running from the head of the St. Lawrence to Georgian Bay, and to a triangular block with lakes Erie and Huron on two sides of it.

The French did not make more than token settlements west of Montreal, being satisfied to populate the valleys of the St. Lawrence and Richelieu. They left the up-country to fur-traders, missionaries and soldiers. With the conquest in 1760 came an influx of British merchants and officials, but they, too, concentrated in Quebec and Montreal and did not alter the French settlement pattern. During the American Revolution, however, United Empire Loyalists began to move into Upper Canada, and after them came other English-speaking immigrants, predominantly American before 1812 and predominantly British after 1815. These settlers filled up the arable belt of present-day Ontario as well as Lower Canada's Eastern Townships. Where they went was dictated on the one hand by natural waterways, essential communications to a pioneer society in heavily wooded country where roads were hard to build. On the other hand, the Shield prevented expansion of agrarian population north and west around Lake Superior into what are now Canada's western provinces. To those responsible for Canadian defence between 1815 and 1855, then, geography presented the two-fold problem of defending a region cut off from direct contact with the sea and stretched thinly along some 500 miles of waterways between Montreal and the western end of Lake Erie, waterways which, in turn, were the northern boundary of a hostile nation.

Canada's social and economic condition heightened this problem. In 1825 the population of Upper Canada was approximately 158,000 while that of Lower Canada was 479,000. By 1850 the respective figures were 791,000 and 840,000, and by 1852 the population of Upper Canada (Canada West after 1841) surpassed Lower Canada (now Canada East). Thus the number of British Canadians, swelled by heavy immigration from the United States and Britain, increased more than four times as fast as the French Canadians, who relied entirely on the cradle. After 1815, therefore, with every passing year there were many more British subjects to be protected in the area of Canada most difficult to defend.

Simultaneously the number of their potential enemies in the adjacent American states to the south increased at about the same rate. This meant that the western states of the Union retained their great population lead over Upper Canada. In 1820 the population of the trans-Appalachian states was approximately 2,602,000 and in 1850 it was some 10,436,000. In terms of the relative manpower ratio, therefore, the Upper Canadians had no chance of defending themselves against attack from the south without heavy British support.

The characteristics of the Canadian people again added to the problem of defence. The French Canadians comprised a homogeneous and conservative society led by a small spiritual and professional hierarchy dedicated to the preservation of their culture—their language, laws, and the Catholic religion—under British rule. They were predominantly agrarian, not commercially-minded or market-oriented. By contrast the English-speaking Canadians were of mixed ancestry, of Loyalist, American, English, Scots and Irish extraction, and of a variety of religious denominations. They were very market-oriented, and the most dynamic element among them were businessmen, not professionals and clergy. Until about the second half of the nineteenth century the most important business leaders were concentrated in Montreal.

These Montreal merchants did much to shape Canada's social and economic development. After 1760 they settled among the French Canadians and took over the management of the fur trade from old rivals displaced by the British conquest. Aside from the codfish of the Grand Banks, primarily a concern of the maritime provinces, beaver fur was Canada's first great staple, the basis of an economy which relied upon a vast network of continental waterways. The Montreal merchants extended their fur empire all the way to the Pacific and Arctic oceans. They did so, however, in the face of increasing opposition from the Hudson's Bay Company, which, with cheaper communications, forced them out of the trade by 1821. But by then the merchants had already developed two other staple exports for the British market, wheat in the 1790s and timber during the first decade of the nineteenth century. A preferential British tariff system aided these enterprises between 1815 and 1846. Like fur, both these staple industries rested on water communications.

Timber and wheat, however, required cheaper transport because they were bulkier with a smaller value relative to weight. For example, until the introduction of steam tugs to pull timber rafts on the lakes in the 1840s, it was uneconomical to ship timber to Montreal from regions more distant than the banks of the St. Lawrence and Ottawa rivers. Wheat and food stuffs could be profitably shipped to Montreal not only from the western regions of Upper Canada but also from the American middle west. But the profit margin was slim unless transportation costs could be reduced by the use of larger vessels—preferably steamers—which could navigate the waters of the St. Lawrence without constant transshipment at rapids. The Montreal merchants and businessmen in Upper Canada knew that if they could create a St. Lawrence seaway by building canals they could increase their profits and economic power. Besides facilitating the flow of Upper Canadian commerce to Britain, such a seaway would also draw the much greater agricultural output of the American middle west to Montreal, where American farmers could take advantage of British preferential tariff regulations applying to Canada. In short, the merchants saw the chance to build a commercial empire by improving east-west communications along the St. Lawrence and north-south communications from the Great Lakes to the middle western states. Potentially, Montreal could become the North American metropolis.

The zeal for canals in Montreal was matched by similar ambitions among merchants in the cities of America's eastern seaboard. The first decades of the nineteenth century were the era of North American canal building. Montreal, Boston, New York, Philadelphia, and Charleston all competed to build cheap waterways to the west in order to tap the increasing output of natural products. The westerners, in turn, worked on canals of their own to feed into those from the eastern seaboard. Because of their greater population and resources, the Americans built canals faster than the Canadians—the most successful American canal builder being De Witt Clinton, under whose direction New York completed the Erie Canal in 1825.[8]

This was a blow to the Montreal merchants. By giving New York access to the middle west, the Erie Canal reduced the natural advantage of the unimproved St. Lawrence route and siphoned both American and Upper Canadian produce away from Montreal. Until they could build their own canal system, the merchants had only the advantage of British preference to counter New York competition. As it turned out, by the time they finished their canals in 1848 their tariff preference was gone because of Britain's adoption of free trade in 1846. By that time, too, railways were replacing canals as the chief commerce carriers, and because here also the Americans were in the lead, New York rather than Montreal became the principal metropolis of the continent. Yet the Canadian merchants did not fail to realize their ambitions for want of trying, and perhaps it was their effort to build a commercial empire along the St. Lawrence-Great Lakes axis which ultimately gave Canada an economy sufficiently independent from that of the United States to form the basis of the present Canadian nation.

But for the Ordnance after 1815 these Canadian efforts constituted the leading problem. The colonists were improving the line of communications most exposed to potential American attack, and were promoting better communications with American states south of the Great Lakes which to the Ordnance were possible invasion routes. Thus the department faced the job of creating a safe alternative waterway to Upper Canada along the Ottawa River and the Rideau Canal. It also attempted to stop projects designed to improve communications between Canada and the United States, and as a result clashed with powerful civilian interests which made Ordnance military tasks more difficult to execute.

Finally, the political evolution of Canada after 1815 brought additional troubles to the department. By means of the Constitutional Act of 1791 British statesmen had attempted to solve two main problems, the conflict between French and English Canadian interests, and the sort of weaknesses in colonial executive government which had led to the American Revolution. After the War of 1812 it became increasingly clear that the act was defective in regard to both aims. The geographic division of Canada along ethnic lines into upper and lower provinces failed to quell Anglo-French antagonism. Furthermore, the strengthening of colonial executives produced oligarchies which dominated British administrators and led to popular demands for majority rule in both provinces. The Colonial Office recognized these problems without providing adequate solutions. In 1837 extremist groups of republican democrats resorted to armed rebellion. The shock effect of their small insurrections destroyed the Constitutional Act and the oligarchies. From 1838 to 1846, pushed by Canadian reformers demanding responsible government on the British cabinet model, and by British reformers insisting

---

8   George Rogers Taylor, *The Transportation Revolution, 1815-1860* (New York, reprinted ed., 1968), pp. 32-55.

on free trade and the removal of old mercantilist reasons for maintaining close control over colonies, the imperial authorities by degrees made Canada into the first major self-governing colony.

Throughout the period of Ordnance activity in Canada, therefore, the colonists exerted a growing influence over their local affairs and the imperial government deferred more and more to their wishes. When the Ordnance tried to stop civilian projects which might endanger provincial security it received little support from senior departments of state. The Canadians had sufficient political strength to ignore or hamper the efforts of a British department which functioned as though it were in a colony under total imperial domination.

The Ordnance had no official or regular channels of communication with the Canadians or their government. Because it was striving to provide the colonists with adequate defences it expected co-operation, but until the 1860s Canadians considered that Britain had a duty to defend them while they occupied themselves with the quest for material progress. Even when Britain paid the bills for defence projects they were impatient when a department like the Ordnance, which never consulted them, tried to carry out policies which seemed to interfere with their civilian objectives.

In contrast with the Ordnance the U.S. army engineers followed the orders of a popularly elected government and directly served the interests of American citizens. These engineers assisted settlers and entrepreneurs by opening new territories for expansion and exploitation. Unlike this moving frontier of the American west, the Canadian frontier was an international boundary threatened by the expansion of ambitious Americans, many of whom wanted the whole North American continent under their flag. Before the War of 1812 Upper Canada to them was merely another portion of their own west, another open area for settlement; by 1812 most Upper Canadians were in fact Americans. After 1815, when Upper Canada began to fill up because its arable land was constricted by the Precambrian Shield, many British immigrants moved through it and joined the flow of population into the American west, which now became the Canadian frontier of settlement.[9] The Turner thesis notwithstanding,[10] then, the Canadian environment between 1815 and 1855 in which the Ordnance worked, did not correspond to the expanding Turnerian frontiers in which the American army engineers carried out their contributions to civilian development. Canada was in a pioneer, underdeveloped state, but Canadians strove to overcome that condition within a geographically limited area which had no western hinterland within its political boundaries until the 1860s. The Ordnance did indeed help the Canadians toward material progress, but mostly in ways which were incidental to its major preoccupation with the safeguarding of the international frontier, the Canadian-American boundary.

Concern over this boundary, and with conditions of the North American environment, had already shaped British military policy toward Canada well before the

---

9   See Fred Landon, *Western Ontario and the American Frontier* (Toronto, 1941).
10   Walter Noble Sage, "Some Aspects of the Frontier in Canadian History," *C.H.A. Report* (1928), 162-72.

Ordnance extended its activities to the colonies. This policy, in turn, influenced the department. The Ordnance took up the task of completing projects begun by other military authorities, projects which must be noted in order to understand and evaluate the department's record.

As early as 1779, when it became obvious that a new British colony was likely to develop in what became Upper Canada as a result of the American Revolution, imperial authorities began to struggle with the problems of military communications. At this stage, while population west of Montreal was small and while the St. Lawrence was still somewhat sheltered from American attack by thick, trackless forests along its southern bank, British soldiers applied their limited resources to the improvement of the most direct water route to Lake Ontario. Between 1779 and 1783 a detachment of Royal Engineers took the first steps to get past some of the biggest obstacles on the St. Lawrence River. They built two small canals at Coteau-du-Lac and the Cedars to facilitate the movement of bateaux around rapids.[11] In December 1800 Colonel Gother Mann, Commanding Royal Engineer, proposed to enlarge these canals and to add a third one at the Cascades. He urged that these improvements were necessary to keep pace with increasing traffic on the river.

The arguments he used to gain support for his project are of interest because they were to be repeated many times by military canal promoters. Like others after him, Mann had to convince a reluctant imperial government to part with large sums of British taxpayers' money; he therefore needed more than military reasons for his project. It would, he stated, also contribute substantially to the civilian development and welfare of Upper Canada, a province with an expanding population and trade requiring improved communications with Britain. As he expressed it,

> The great convenience of the Locks, when in a proper state in avoiding the dangerous rapids, and in facilitating the navigation of the River, are objects of so much consequence both to Government, and to the Trade of Upper Canada, that it seems requisite they should be immediately attended to: the Toll arising from them will fully justify and repay the expence of any reasonable and substantial repairs and improvements which may be required.[12]

Mann's emphasis on tolls was a particularly important point. The British government would not only make valuable imperial improvements by building canals, but these canals would pay for themselves and ultimately cost nothing. This argument was used repeatedly in support of later military communications projects.

Mann's proposal was approved, and London ordered him to start construction in June 1801.[13] Captain R. H. Bruyeres, R.E., assisted from 1802 by Lieutenant John By, R.E., took charge of the construction work. Mann was confident that these officers

---

11 R.G. 8, C38, pp. 1-18, Mann to Hunter, Dec. 24, 1800.
12 Ibid.
13 R.G.8, C620, pp. 8-9, Portland to Hunter, June 6, 1801.

could keep costs within the original estimates and that tolls would repay all the money spent.[14] In 1804, however, Bruyeres indicated that costs were rising beyond expectations because of difficulties with rock excavation and an inadequate labour supply.[15] Nevertheless the work went on, and in October 1806 By—now promoted to Second Captain—completed the canal at the Cascades.[16] Although additional improvements were made in later years, sometimes by local civilians like David Sheek of Cornwall,[17] the Cascades project marked the completion of the first military canal system in Canada. The canals themselves passed under the management of the Commissariat Department.

John By's participation in this effort is important. During his service in Canada between 1802 and 1811 he acquired the experience and reputation which later led to his appointment as builder of the Rideau Canal. He apparently made a favourable impression on his superiors. His immediate commander, Captain Bruyeres, felt that he was ". . . well acquainted with the Work, and has conducted it with great judgment and ability."[18] As it happened, however, his performance on the small Cascades project proved to be a doubtful index of his ability to direct the larger Rideau work.

The Commissariat canals served their purpose until 1812 but could not meet the demands imposed by wartime conditions. In 1814 W. H. Robinson, the Commissary General, urged the commander-in-chief, Sir George Prevost, to supplement them by a canal from Montreal to Lachine and by other works on the upper reaches of the St. Lawrence. Without such improvements, Robinson argued, the already staggering cost and difficulty of providing British forces in Upper Canada with essential supplies could increase to the point where, with constant intensification of operations expected in 1815, it would be impossible to keep up sufficient military and naval strength to hold the province.[19] Besides slowness and unreliability, transportation costs along this sole military waterway were almost prohibitive. By one estimation, it cost £200 to haul a 24-pounder canon to Kingston.[20] Approximately £630,000 was spent on military transport between Quebec and Upper Canada during the three years of the war.[21]

Furthermore, the Commissariat canals were extremely vulnerable to American attack. If they were destroyed and the St. Lawrence route was closed, the British forces in Upper Canada could be starved into surrender. The Americans were slow to grasp this fact until 1814, when they made plans to cut the river route the following year; the Treaty of Ghent brought peace before they could act. In the meantime, the British learned of the intended assault. Apparently espionage agents working under the

---

14 R.G.8, C38, pp. 45-50, Mann to Hunter, Aug. 31, 1802.
15 Ibid., pp. 67-69, Bruyeres to Green, Jan. 16, 1804.
16 W.O. 55/858, p. 514, Bruyeres to Morse, Oct. 25, 1806.
17 R.G.8, C38, p. 41. Sheek to Hunter, Mar. 27, 1802.
18 Ibid., p. 69, Bruyeres to Green, Jan. 16, 1804.
19 C.O. 42/157, pp. 355-57, Robinson to Prevost, Nov. 14, 1814.
20 See W.O.44/32, pp. 157-59, By to Smyth, Dec. 10, 1827, W.O.55/864, pp. 63-64, Byham to Mann, July 3, 1827, and Legget, *Rideau Waterway*, p. 25.
21 W.O.46/136, pp. 58-64, Thomas to Byham, Apr. 10, 1840.

Canadian, Colonel "Red George" Macdonell, reported details of the plan in 1814,[22] and after the end of hostilities Major General Jacob Brown of the U.S. Army disclosed it to Major General Sir Frederick Robinson.[23] British officers therefore knew that in the next war—expected by both sides because of the inconclusive results obtained at the end of 1814—the Americans would close the St. Lawrence route. A safe alternative well away from the international boundary was now essential to the security of Upper Canada. Thus from 1814 British soldiers turned to the Ottawa-Rideau route; the need for a waterway along this line was probably the chief lesson learned from the War of 1812 by those responsible for Canada's defence.

The war exerted a strong effect on the thinking of American as well as British soldiers and political leaders. It ended in stalemate because neither side could exploit its advantages adequately. Britain commanded the seas and the United States had no real defences against the Royal Navy. Britain had the capacity to assault American maritime commerce and seaboard territory with virtual impunity. But until early 1814 the Royal Navy was occupied with the more pressing task of crushing Napoleon's empire. It applied its full force against the Americans only during the last months of the war. In turn, the Americans had the upper hand in the interior of the continent; here their greater manpower, material resources and better communications gave them effective superiority over the Canadian and British troops west of Montreal. Because of poor generalship and organization, however, they made a series of mistakes and failed to concentrate their strength against the decisive weak points in the British lines, particularly the all-important St. Lawrence communication. In the next war leaders on both sides planned to exploit their advantages properly and to take defensive measures which might counter their disadvantages. A series of international crises stemming from the Canadian rebellions of 1837, the boundary dispute between Maine and New Brunswick and the 1845 controversy over Oregon kept the war planners at work. Diplomacy solved each crisis, but in the interim admirals and generals exerted themselves in improving their respective offensive and defensive positions.

The United States began a major programme of fortification along its Atlantic coasts in March 1815, concentrating on projects like Fortress Monroe in Hampton Roads. The main aim was to counter British naval superiority.[24] The Americans were relatively unconcerned about defensive works along the British North American boundary, being confident that here they could defeat the British by vigourous offensive action.[25] While they knew that a successful assault on Upper Canada and Montreal would depend on control of the Great Lakes and Lake Champlain, they also believed that

22 George Raudzens, " 'Red George' Macdonell, Military Saviour of Upper Canada?" *Ontario History*, LXII (Dec. 1970), 199-212.

23 C. P. Stacey, "An American Plan for a Canadian Campaign. Secretary James Monroe to Major General Jacob Brown, February 1815," *A.H.R.*, XLVI, 2 (Jan. 1941), 348-58.

24 Bourne, *Britain and the Balance of Power*, pp. 50-51.

25 Ibid., p. 44.

on inland waters they could build navies superior to British fleets. The Rush-Bagot agreement of 1817, establishing all but complete disarmament on the border lakes, was to their advantage. With their greater local resources they could easily outstrip the British in a naval rearmament race from an equal start.[26] Thus, while Congress voted funds for Atlantic coast defences, American generals made contingency plans for overland offensives into Canada.

The British, in turn, were confident that their naval squadrons from the major bases of Halifax and Bermuda could win and maintain the advantage along the Atlantic coast. At the same time they were most sensitive to the vulnerability of their Canadian possessions west of Montreal. In terms of broad strategy it was more logical to evacuate Upper Canada in wartime and to punish the Americans so severely on the seas and along their coasts as to force them to return the lost province at the peace settlement. But the surrender of Upper Canada, even temporarily, was politically unacceptable. Despite the fact that the task looked nearly hopeless the imperial government felt an obligation to defend its loyal Canadian subjects. The Duke of Wellington summarized this British view in 1825: "It it impossible for His Majesty's Government to withdraw from these Dominions. Whether valuable or otherwise, which can scarcely be a question, the Honor of the Country would require that they should be defended in War. . . ."[27] The soldiers therefore had to do their best to prevent a successful American invasion. With few exceptions they agreed that while offence might be the best form of defence, in Canada all they could hope to do was to hold the ground.[28] Even to follow a purely defensive strategy, however, they needed better and safer military communications beyond Montreal, protected by massive fixed fortifications.[29]

Out of these considerations emerged a standardized scheme for Canadian defence. On November 7, 1818, Lieutenant Colonel John Harvey, a War of 1812 veteran and deputy adjutant general to the governor-in-chief, the Duke of Richmond, sketched what was probably the first full outline of this plan. In his "Memorandum on the Defence of the Canadas" he urged that the St. Lawrence waterway should be abandoned for an alternative route between Montreal and Kingston using the Ottawa and Rideau Rivers, and that the two terminals should be heavily fortified. To make sure that British sea power had secure access to Montreal, the citadel at Quebec should be much strengthened. Beyond Kingston Harvey urged that the Niagara peninsula and points west to the confluence of Lakes Erie and Huron should also be fortified despite the probability that it might be impossible to supply and reinforce garrisons in the western regions of Upper Canada during hostilities. The people of these regions had to have

---

26   Ibid., p. 24.
27   C.O.42/205, p. 193. Wellington to Bathurst, Dec. 6, 1825.
28   Hitsman, *Safeguarding Canada*, pp. 121-29.
29   The British government committed itself to defend Canada almost exclusively by land forces after the War of 1812. Various proposals to build naval forces on the lakes as a first line of defence received little support. By 1838 the Royal Navy establishment in Upper Canada was almost nonexistent. See Bourne, *Britain and the Balance of Power*, pp. 4, 23, 32.

visible assurance that Britain meant to protect them. Only then would they remain loyal and serve as militia auxiliaries to the regular troops in wartime.[30]

Three days after Harvey drew up this document his commanding officer, the Duke of Richmond, wrote a more elaborate version of it and sent it off to the colonial secretary, Lord Bathurst.[31] The latter passed it on to the master general of the Ordnance, Wellington, who returned a revised version on March 1, 1819.[32] This document, with minor changes from time to time, became the official Canadian defence plan for the next two-and-a-half decades.[33] In 1840 Lieutenant General Sir Richard Jackson referred to it in a memorandum on Canadian fortifications stating that "the nearer we can make our measures approximate to those recommended by the Duke the better."[34]

By 1819, then, the necessary master plan was ready and accepted; but its execution was another matter. The British Parliament and Treasury, growing more concerned about economy than colonial security, were unwilling to provide the money and those directly responsible for Canadian defence had to battle for every penny. It took them until the late 1840s to complete even a part of all the fixed fortifications stipulated as indispensable by Wellington. Aside from military canals, the Duke of Richmond began the construction of Montreal's southern outwork, Fort Lennox, in 1819. This fort was never completed and the local security of Montreal remained a constant worry until the 1860s. The citadel of Quebec, begun in 1820, was completed in 1831.[35] It was the only important Canadian defence work considered adequate by military experts. The Kingston fortification project, started in 1826, was not completed until after 1846, and the finished product was unsatisfactory.[36] Other works of lesser importance such as barracks and repairs or alterations of fortifications already in existence by 1819 were fully or partially completed by about 1846 and the all-important Ottawa-Rideau canal

---

30   G.D.45, Section 3 (5) 332, "Memorandum on the Defence of the Canadas," by Lieutenant Colonel John Harvey, Nov. 7, 1818.

31   C.O.42/179, pp. 119-22, Richmond to Bathurst, Nov. 10, 1818. It is probable that Richmond's plan was derived from Harvey's. In 1824 Lord Dalhousie cited Harvey as the leading expert on Canadian defence problems. See G.D.45/3/354, Dalhousie to Taylor, May 20, 1824.

32   G.D.45/3/335, Wellington to Bathurst, Mar. 1, 1819.

33   One historian argues that Richmond's plan contained nothing new. John Graves Simcoe, Sir James Craig and Sir George Prevost had expressed the same ideas earlier (see Hitsman, *Safeguarding Canada*, p. 117). But neither Simcoe nor Craig had contemplated an Ottawa-Rideau canal system. While Prevost did propose such a system, he failed to draw up a complete Canadian defence plan incorporating it as did Harvey, Richmond and Wellington.

34   W.O.1/536, p. 140, "Memorandum upon the Canadian Frontiers" by Lieutenant General Sir Richard Jackson, Nov. 1840.

35   C. P. Stacey, "A Note on the Citadel of Quebec," *C.H.R.,* XXIX (Dec. 1948), 389-90.

36   Both the fortresses at Quebec and Kingston were major works but they caused the Ordnance no undue problems. Both were financed only at the speed with which Parliament allocated funds and the debate over money took place at the highest levels of government, with departmental officers acting only as advisors when requested. For example, see Bourne, *Britain and the Balance of Power*, pp. 164-67. Neither project involved undue construction difficulties or technical innovations. A detailed examination of the way these forts were built does not add a great deal to an understanding of the department's role in Canada.

system was in operation by 1834. By 1851 the Ordnance controlled 15 garrisoned fortresses or stations along the 720 mile frontier between Quebec and Amherstburg (Quebec, Sorel, Montreal, Chambly, Saint John, Isle Aux Noix, Bytown, Prescott, Brockville, Toronto, Niagara, London, Amherstburg, and Penetanguishine); excluding Royal Engineers, Artillery and Sappers, 282 departmental "civil officers" and civilian employees executed the required day-to-day maintenance work.[37] This was a large establishment and a heavy burden to the department, but it was not the complete, integrated system of defence which Wellington and other senior officers regarded as the only sure means of preventing American conquest.[38]

All these projects, including the canals, were thus started before the Ordnance took up its full range of Canadian duties. The department assumed responsibility for completing them and to that extent the failure to implement the whole plan belonged to it. The plan was largely determined by the conditions of the Canadian environment which hampered all Ordnance activities.

Aside from these problems, however, weaknesses within the Ordnance and the whole imperial administration also contributed to the department's shortcomings. Like other branches of the government it had difficulty controlling its overseas agents and its financial arrangements were so confusing they tended to make the already niggardly Treasury increasingly reluctant to approve defence spending outside the British Isles. In the case of the vital Ottawa-Rideau project these defects became especially pronounced, to the point where they jeopardized all future defence efforts. From its inception, well before the Ordnance became involved, the military canal scheme was instigated by imperial officers too often acting on their own unauthorized initiative, and able to carry on only by exploiting London's inadequate knowledge of the details about colonial expenditures.

---

37 W.O. /564, pp. 299-308, "Report of the Committee appointed by the Lords Commissioners of Her Majesty's Treasury, to inquire into the Establishment and Expenditure of the Naval, Ordnance, and Commissariat Departments in the Colonies and Foreign Possessions of the Crown," July 22, 1851.

38 Bourne, *Britain and the Balance of Power*, pp. 36-52, and Hitsman, *Safeguarding Canada*, pp. 120-29.

Chapter IV

The Struggle for Military Canals, 1814-1825

By 1825, when the Ordnance took up its primary canal building task, the Rideau Canal project was the last link necessary to complete the Ottawa-Rideau waterway. It was because the other works were already being constructed that the imperial government approved this final, most difficult part of the plan. Approval was granted because of the sustained efforts of senior military officers with Canadian responsibilities. In the face of strong resistance from the Treasury, promoters of safe military communications had since 1814 conducted a relentless struggle which heralded the later problems of the Ordnance.[1]

The origins of the Rideau project have been traced, somewhat tenuously, back to the late eighteenth century. Historians have noted that between 1814 and 1825 various efforts were made to begin construction but that nothing substantial was achieved.[2] In fact the local agents of empire, Sir George Prevost, Gordon Drummond, Sir John Sherbrooke, the Duke of Richmond and Lord Dalhousie—aided by Wellington and Lord Bathurst in London—gained much through persistent effort. Until 1816 they attempted to start the whole Ottawa-Rideau project as a single work, but the Treasury vetoed their scheme on the grounds of cost. Therefore they changed their tactics and began to build the waterway piecemeal. The Canadian merchants were just as eager for

---

1 The substance of this chapter first appeared as an article in the *Canadian Historical Review*. George K. Raudzens, "The Military Impact on Canadian Canals, 1815-25," *C.H.R.*, LLV (Sept. 1973), 273-86.

2 The main historical accounts of the Rideau Canal are H. P. Hill's "The Construction of the Rideau Canal, 1826-1832," *O.H.S.P.R.*, XXII (1925), 117-24; H. P. Hill's "Lieutenant Colonel John By, a Biography," *Royal Engineers Journal*, XLVI (Dec. 1932), 522-25; Robert Legget's *Rideau Waterway* and Robert Brown Sneyd's "The Role of the Rideau Waterway, 1826-1856" (University of Toronto M.A. Thesis, Sept., 1965). Legget's is the most comprehensive of these accounts, but even he passes over the 1814 to 1825 period rapidly and conveys the impression that British officers did not do much to promote the Ottawa-Rideau scheme during these years. Sneyd's work is the only properly documented and scholarly attempt, but he treats the period before 1825 briefly.

civilian canals and equally short of funds. If they were to compete successfully with the wealthier Americans they would need imperial aid. In these circumstances the soldiers and merchants for a time combined their interests; if the military could show that defence expenditure would advance Canadian (and imperial) prosperity, and if the businessmen could thus gain funds to make up what they could not attract from private sources, both would benefit. Therefore both appealed to London accordingly, and both succeeded in part. But the soldiers gained the most because their influence on government was more direct. They pointed out that the Canadians were willing to spend large sums for canals leading to Upper Canada. Would the Treasury contribute just enough to adapt the civilian canals to military needs? When this small grant was allowed, the soldiers extracted gradually more and more, pushing works up the Ottawa on the pretext the Canadians wished to open this route rather than the St. Lawrence. The civilians did get some of the imperial funds for their St. Lawrence seaway, but the soldiers diverted most of the money up the Ottawa. While this was taking place, New York completed the Erie Canal. Then, in 1825, the military promoters asked for the Rideau Canal to complete their system, for by then all the money spent on the Ottawa would be wasted without it. By these often devious means they attempted to safeguard the empire irrespective of the wishes of the colonists and in defiance of high policy in London.

As has been noted, the initial British wartime transportation crisis turned officers toward schemes to improve the St. Lawrence route. In June 1814 Lieutenant General Gordon Drummond, administrator of Upper Canada, urged that the St. Lawrence military canals should be enlarged.[3] Commissary General W. H. Robinson proposed a similar project the following November. In particular he wanted a canal to bypass the Lachine Rapids at Montreal and requested that a "Scientific Director, acquainted with Canal Making in England" should be brought to Canada as soon as possible to start this work.[4] Robinson's Lachine canal would eliminate a major barrier to navigation impeding access to both the St. Lawrence and Ottawa Rivers.

Then, in December 1814, Lieutenant Colonel "Red George" Macdonell put forward the idea of building a military route along the Ottawa and Rideau waterways. Early in 1813 his Canadian light infantry and militia had driven the American garrison out of Ogdensburg, New York. He had launched this assault on his own initiative, in an effort to push the enemy away from the St. Lawrence supply line to Upper Canada. Because of his concern for the security of that line he also began to search for an alternative route, personally explored the Rideau river and lake system, and concluded that it could be made navigable for use in conjunction with the Ottawa. When his espionage agents reported the American plan for a major assault on the St. Lawrence line in 1815, he proposed his alternative route to Drummond and the governor-in-chief, Prevost. These men put their full support behind his plan. From this point forward,

3 R.G.8, C38, pp. 85-86, Drummond to Prevost, June 16, 1814.
4 C.O.42/157, pp. 355-57, Robinson to Prevost, Nov. 14, 1814.

therefore, the soldiers linked Robinson's Lachine project with Macdonell's proposal, thus launching the campaign which culminated in the building of the Rideau Canal.[5]

In February 1815 deputy commissary general I. W. Clarke suggested to Robinson that while they waited for the scientific adviser from England they should survey the proposed Lachine line and make everything ready for the start of construction as soon as London sent approval.[6] A few days later Baron Francis de Rottenburg, administrator of Lower Canada, assigned an officer to conduct the survey,[7] and simultaneously Prevost took action at higher levels. In a letter to Lord Bathurst on February 23, 1815, he stated that he had submitted the Lachine plan to the Lower Canadian legislature. The Canadians, eager to improve civilian communications along the St. Lawrence, had been so enthusiastic they had voted to contribute £25,000 if the imperial government undertook the project. Prevost urged London to take full advantage of this vote and to send Robinson's scientific expert as soon as possible.[8]

Meanwhile the Canadians decided that London was sure to build the canal and passed an act placing £25,000 at the disposal of the governor-in-chief.[9] This sum proved to be the wedge with which the military canal promoters pried imperial funds out of the Treasury. Drummond, now administrator of Lower Canada, took the next step in the campaign, probably without realizing fully what he did. In June 1815 he asked Bathurst for the services of a detachment of the Royal Staff Corps.[10] Governor Sir John Sherbrooke of Nova Scotia had informed him of the arrival of 121 officers and men of this corps of military engineers at Halifax, where they were not immediately needed. Could Drummond use them? Writing to Bathurst, Drummond stated that he did indeed have work for them, not only on the proposed Lachine Canal but also on the improvement of water communications between Montreal and Kingston via the Ottawa-Rideau route.[11]

Thus Drummond did two things. First, he requested the use of a military work force which, once assigned to canal work could be directed under the independent authority of officers on the spot. These officers would therefore have the means to undertake projects difficult to stop by the home government. The Staff Corps was used in precisely this way to start building canals on the Ottawa section of the Ottawa-Rideau scheme. Second, Drummond tied this scheme together with the Lachine project. He therefore made it possible for himself and his successors to use the Canadian sum of £25,000 as a device with which to push the imperial government into accepting the whole military canal plan.

---

5 Raudzens, " 'Red George' Macdonell, Military Saviour of Upper Canada?"

6 R.G.8, C38, pp. 108-09, Clarke to Robinson, Feb. 14, 1815.

7 Ibid., p. 113, De Rottenburg to Prevost's military secretary, Feb. 20, 1815.

8 C.O.42/161, pp. 69-70, Prevost to Bathurst, Feb. 23, 1815.

9 Glazebrook, _History of Transportation in Canada_, I, 75.

10 The Staff Corps was created by the Duke of York in 1799 as a rival organization to the Royal Engineers. See Glover, _Peninsular Preparation_, p. 102.

11 C.O.42/162, Drummond to Bathurst, June 8, 1815.

As the reply from the Horse Guards concerning this request for the Staff Corps indicated, Drummond's manipulation took effect quickly. Sir Henry Torrens wrote to him on July 24, 1815, stating that, in view of the importance of improved military water communications between the two Canadian provinces, the Staff corps detachment could be used for canal building as Drummond wanted.[12] Drummond also won the support of the colonial secretary. On October 10, 1815, Bathurst wrote to him that the imperial government was "deeply impressed" with the need for an interprovincial military communication. Drummond was therefore to make surveys, plans and cost estimates, so that London could decide whether to commence the Lachine and Ottawa-Rideau projects separately or together.[13]

Meanwhile the project gained support from another direction. The British government was carrying out a plan to locate military settlements in strategic areas of the Canadas; the aim was to dispose of superfluous soldiers and to increase the population of the provinces with veterans who could strengthen militia forces in emergencies.[14] On July 29 Major General Sir Frederick Robinson reported to Bathurst that one of the most important places for concentrating the soldier settlers was the Rideau River, "lest in the event of another War with the United States it should be deemed eligible to establish a communication between Montreal and Kingston. . . ." But to build this communication in wartime would be most difficult, warned Robinson. It was best to do this job as soon as possible, before a resumption of hostilities. He had himself already taken steps to draw up plans and estimates. Once the Rideau area was settled by veterans it would have man-power, food, forage, horses and bullocks, all of which would facilitate rapid canal construction. In turn the canal would provide the settlers with an outlet for their produce and increase their prosperity.[15] In other words the military settlements and the canals were complementary, each facilitating the other. The promoters of the Ottawa-Rideau project made good use of this argument.

Official approval came quickly as a result. Apparently the £25,000 voted by the Canadians impressed London because early in 1816 the Lords Commissioners of His Majesty's Treasury agreed to provide financial aid for the Lachine Canal project. If the Canadians paid half, the Treasury was prepared to authorize payment of the rest out of the Army Extraordinaries fund.[16] The Lords Commissioners, however, made no provision for the other parts of the military waterway. They were not yet ready to declare themselves either way on the larger undertaking.

---

12 R.G.8, C38, p. 121, Torrens to Drummond, July 24, 1815.
13 Ibid., Bathurst to Drummond, Oct. 10, 1815.
14 See Helen I. Cowan, *British Emigration to North America, 1783-1837* (Toronto, 1928), pp. 65-95; Robert England, "Disbanded and Discharged Soldiers in Canada Prior to 1914," *C.H.R.*, XXVII (1964), 1-18; Norman Macdonald, *Canada, 1763-1841: Immigration and Settlement: The Administration of the Imperial Land Regulations*, pp. 39-67; and George F. Playter, "An Account of Three Military Settlements in Eastern Ontario—Perth, Lanark and Richmond, 1815-20," *O.H.S.P.R.*, XX (1923), 98-104.
15 C.O.42/356, pp. 69-72, Robinson to Bathurst, July 29, 1815.
16 C.O.42/180, G. Manning to Goulburn, Jan. 6, 1816.

But the colonial secretary was ready to proceed. On January 8, 1816, Bathurst wrote to Lieutenant Governor Gore of Upper Canada stating that as a result of a number of communications from officers and governors in Canada it was obvious all men in positions of military authority there were virtually unanimous about the immediate need for the Ottawa-Rideau system. Therefore Gore was to follow Robinson's plan of settling old soldiers on the Rideau River and also to advertise for tenders from civilian contractors for the construction of all the canals.[17]

While this despatch was crossing the Atlantic, Gordon Drummond was already taking action. He ordered parties of Royal Engineers to explore the Ottawa and Rideau river systems.[18] Lieutenant J. Jebb, R.E., went up the Rideau route from Kingston in April[19] and Lieutenant R. J. Baron, R.E., went up the Ottawa from Montreal in May. Both were instructed to make detailed surveys, draw up large-scale maps, note all obstacles to navigation and work out plans for canals and locks. In both cases they were to seek routes sufficient for bateaux only.[20] As Jebb began to send in reports to the Royal Engineer headquarters at Quebec,[21] Drummond appointed Captain Samuel Romilly, R.E., as supervisor, with the duty to co-ordinate surveys, plans and cost estimates for the whole route. Romilly was still at work in July[22] when Francis Gore apparently decided that enough preliminary information had been collected. On July 7 he asked Bathurst's permission to advertise for Rideau Canal tenders;[23] the advertisements appeared in February 1817. A notice published by the lieutenant governor's office of Upper Canada stated that tenders would be received until June 30 for the building of a waterway from Lachine to Kingston via the Rideau, to take boats with a draught of two feet and between three feet ten inches and twelve feet in width.[24]

The promoters were thus rushing their government into a major project without so much as completed plans and estimates. Given the slowness of communications and bureaucratic action, such speed indicated their sense of urgency. But fast action could not obviate the need for careful organization and regard for parliamentary sensibilities about expenditures in colonies.

When the Treasury belatedly received news of the actions taken with Bathurst's support it at once vetoed the project. On October 15, 1816, while officers in Canada were proceeding on the basis of Bathurst's authorization—at least four months before news of the veto reached them—A. W. Hawkes from the Treasury wrote to Henry Goulburn at the Colonial Office as follows:

---

17 R.G.7, G1, Vol. 58, pp. 4-5, Bathurst to Gore, Jan. 8, 1816.
18 C.O.42/166, p. 17, Drummond to Bathurst, Jan. 22, 1816.
19 R.G.8, C38, p. 139, G. Nicholls to Jebb, Apr. 27, 1816.
20 Ibid., p. 159, Nicholls to Baron, May 9, 1816.
21 R.G.8, C40, pp. 157-97, Jebb to Nicholls, June 8, 1816, and C.O.42/166, pp. 350-54, Jebb to Durnford, June 1816.
22 C.O.42/167, p. 29, Sherbrooke to Bathurst, July 23, 1816.
23 C.O.42/357, p. 248, Gore to Bathurst, July 7, 1816.
24 R.G.8, C39, p. 11, Notice issued by the lieutenant governor's office of Upper Canada at York, Feb. 19, 1817.

Having laid before the Lords Commissioners of His Majesty's Treasury the voluminous and detailed reports (transmitted by you) as to the best means of defending His Majesty's Provinces in North America and facilitating the transport of Stores, I have it in command to return the original papers to you as requested and I am to acquaint you for the information of the Earl Bathurst that my Lords have had this Subject under their attentive Consideration and are fully aware of the great importance of several of the propositions submitted, but, adverting to the expence which would be necessarily incurred in the execution of these Works and the imperfect information which the Papers afford in respect to some of them and also to the circumstances that none of them appear to be of such urgent and immediate necessity that they may not be dispensed with at least for the present, My Lords do not think it would be expedient now to undertake these works when it is necessary that every practicable reduction should be affected in the Public Expenditure [because of heavy public debts incurred during the preceding war].[25]

It seemed, then, that higher authorities in the imperial government had put an effective stop to military canal building in Canada.

Yet the promoters of the project refused to yield. They set out to circumvent Treasury opposition by building the canal system piecemeal. First they took full advantage of the pledge made by the Lords Commissioners to pay half the costs of the Lachine Canal if the Canadians paid the rest. On April 1, 1817, while Captain Romilly was drawing up plans and estimates for the Lachine project,[26] Sir John Sherbrooke, now governor of Canada, urged Bathurst that the proposed cost-sharing agreement for this canal should be implemented immediately. For this a civil engineer from Britain was essential. If one was sent, Sherbrooke stated that Lower Canada would probably provide a sum additional to the £25,000 already appropriated. In fact the Lower Canadians were so eager for the canal they might even pay the full cost after obtaining the services of a civil engineer. The House of Assembly considered that Sherbrooke's predecessors, Prevost and Drummond, had bound the imperial government to provide such a technical expert.[27]

This request was calculated, deliberately or not, to appeal to the Treasury. Sherbrooke raised the possibility that the Treasury might not have to pay any of the money it had promised. All that was now needed was the expense of sending one civil engineer. The Lords Commissioners responded quickly. Only three months after Sherbrooke sent his letter from Quebec they were pleased "to authorize the expense of sending out a competent civil engineer to superintend the undertaking, during such a period as may be necessary to ensure its successful progress. . . ."[28] This small commitment was the first step in the gradual involvement of the imperial government with

---

25 C.O.42/169, p. 205, Hawkes to Goulburn, Oct. 15, 1816.
26 P.P. 1830-31 (135) IX, 15, "Canada Canal Communication," pp. 8-10, Report on the Lachine Canal, by Captain Romilly, 1817.
27 W.O.44/19, p. 8, Sherbrooke to Bathurst, Apr. 1, 1817.
28 Ibid., pp. 8-9, Treasury Minute, July 4, 1817.

military canals on the Ottawa River. The second step, despite Sherbrooke's assurance that a British contribution would probably be unnecessary, was to commit the Treasury to substantial expenditures on the basis of its promise to share Lachine Canal costs.

At this point Captain Romilly completed the first detailed cost estimates for this canal, and their modest size probably helped the promoters to take the second step. Romilly calculated the expense at only £13,780, dividing this sum into military costs of £5,138 and civil costs of £8,642. To the main canal he added two branch canals, which would cost an extra £28,433 and £17,690 respectively. Most of the cost of these he stipulated as civilian. The military cost of the first branch he put at a little over £300, and of the second at just under £4,000. The total project with the branches would involve an expenditure of about £73,000.[29]

Romilly's estimates are not precisely dated but it is certain Sherbrooke used them when he composed a despatch to Bathurst on September 5, 1817. But Sherbrooke's version of the figures contained important changes. For the canal proper he copied out Romilly's sums exactly. After that, however, he included only the estimate for one of the two branches, and put its total cost, not just the £4,000 calculated by Romilly, under the military heading.[30] It is impossible to determine from surviving documents whether Sherbrooke made an honest mistake or deliberately distorted the figures to make total costs smaller than Romilly's and to show that the military expenses surpassed the civilian. In any case the Treasury was bound to infer that if it paid fifty per cent on military grounds, and if military costs far exceeded civil costs, then it was to the advantage of the imperial government to contribute its promised half and let the Canadians pay a large part of the military expenses out of their half. In these circumstances Sherbrooke's version of the figures seems suspect.

Bathurst apparently passed these calculations to the Treasury and used them as an effective argument. On November 20, 1817, Goulburn, his secretary, addressed a strongly-worded letter to the Lords Commissioners reminding them of the long series of reports stressing the vital need for improved military communications in Canada. Lord Bathurst now asked them to act only on the Lachine Canal, a small but most important part of the whole scheme. Goulburn gave assurances that the Canadians would pay their half. He added that the Lachine project would also provide jobs for numerous destitute emigrants recently sent to Canada, who would otherwise have to be supported by the government or be left to starve.[31]

The Treasury responded after deliberating only a little over a month. Basing its decision on Bathurst's recommendation and Sherbrooke's misleading figures, it officially approved the Lachine Canal arrangement.[32] On April 3, 1818, Bathurst wrote to

---

29 P.P. 1830-31 (135) IX, 15, "Canada Canal Communication," pp. 9-10, Three estimates of Lachine Canal costs by Captain Romilly, 1817.

30 C.O. 42/175, pp. 7-12, Sherbrooke to Bathurst, Sept. 5, 1817.

31 W.O. 44/19, p. 9, Goulburn to Harrison, Nov. 20, 1817.

32 P.P. 1830-31 (135) IX, 15, "Canada Canal Communication," p. 11, Treasury Minute, Dec. 30, 1817.

Sherbrooke that the Lords Commissioners had authorized the payment of half the canal costs out of the Army Extraordinaries fund. Sherbrooke was to complete the agreement with the Lower Canadians and "to appropriate the sum already voted by them for this purpose."[33] Thus the second step in the military canal scheme was now completed.

The third step was to extract Treasury approval for an extension of the Lachine project up the Ottawa River, to divert Canadian canal building away from the St. Lawrence route. At this point Lieutenant Colonel John Harvey, Richmond and Wellington applied considerable pressure on their government. All three urged the rapid construction of the whole Ottawa-Rideau waterway despite the Treasury veto of October 1816. Of the memoranda written by these men, Wellington's was the most important in the long run, but the Duke did not immediately follow it up. Between 1819 and 1825 he was too involved with the rebuilding of the Ordnance and political affairs at home to pay much attention to Canada. It was the man on the spot, Richmond, who took action on his own and Harvey's recommendations while official approval was still pending. He authorized the immediate commencement of military canal construction on the Ottawa River.

Even before writing his memorandum of November 1818, Richmond put the Staff Corps obtained by Drummond to work on the planning stages of the Ottawa project. In October 1818, acting on his own authority, the Duke sent Captain J. W. Mann of the Staff Corps to survey the Carillon and Long Sault rapids.[34] In reporting this action to Bathurst, Richmond smoothly combined the Lachine and Ottawa projects for the first time in official correspondence, as though it was self-evident that the two were logically inseparable and always had been. He made this assertion at a time when the Treasury had no idea it had agreed to contribute funds to anything but the Lachine Canal. From Mann's report, stated Richmond,

> . . . it appears evident, that all the difficulties which at present obstruct the navigation of the Ottawa may be overcome, at a trouble and expense very trifling compared with the importance and utility of the object to be attained, an importance of which the people of this province are so convinced, that I have little doubt one half at least of the proposed expense will be cheerfully defrayed by themselves.
>
> If your Lordship concurs with me in this view, I propose employing the Staff Corps on the Ottawa as soon as the weather will admit next spring, and joining to them the assistance of such labourers as the sum appropriated to the object and the country itself will admit of our procuring.[35]

---

33 R.G.8, C39, pp. 32-33, Bathurst to Sherbrooke, Apr. 3, 1818.

34 W.O.44/19, p. 11, Richmond to Bathurst, Jan. 14, 1819, and p. 12, "Report on the Navigation of the *Ottawa* or *Grand River*, . . . Surveyed in October 1818, by Captain J. W. Mann." These documents also appear in P.P. 1830-31 (135) IX, 15, "Canada Canal Communication," pp. 11-12. Captain Mann estimated that the cost of improving the Ottawa River for the navigation of "Batteaux, Gun Boats, Etc. . . ." would amount to £16,740.

35 W.O.44/19, p. 11, Richmond to Bathurst, Jan. 14, 1819.

Richmond thus added the Ottawa canals to the Lachine agreement as though both the Treasury and the Canadians had originally combined them in this way.

There is no evidence that Canadians had strong desires for Ottawa canals.[36] To the dominant Montreal merchant group, for practical purposes of commerce the Ottawa River at this stage led only to sources of timber, and timber rafts could float down to the St. Lawrence without the aid of canals. Yet Richmond said these merchants wanted such canals and were willing to pay half the cost. This was surely a misrepresentation. Furthermore, Richmond had already put the Staff Corps to work on the Ottawa project. To add further weight to his request for approval, he included a report from Lieutenant Colonel Cockburn, deputy quartermaster general, who wanted improved water communications for the military settlers recently placed in the vicinity of the Rideau River.[37]

The Treasury accepted these arguments and Mann's reasonable estimates of costs. In a minute dated May 25, 1819, the Lords Commissioners stated that they would "not object" to a project designed to link Upper and Lower Canada by a waterway along the Ottawa on condition that the legislatures of both provinces paid half the costs.[38] Thus, lacking correct information, the Treasury was prepared to fall in with Richmond's scheme. The Duke had given the Lords Commissioners the impression that canals at Carillon and Long Sault would join up the two provinces, for a total sum of only £16,700. The British government would have to pay only half this sum because both provinces would pay the rest. No one, it seems, had supplied the Lords Commissioners with a map, or they might have had a better idea of what two canals on the Ottawa by themselves could accomplish. Furthermore, Richmond had made no effort to extract promises of financial aid from the Upper Canadians. The Treasury therefore committed itself on the basis of a series of erroneous assumptions supplied by the local imperial proconsul.[39]

On May 20, 1819, Richmond followed up with another despatch calculated to reinforce the misleading impressions he had already conveyed. Reporting on his efforts to get Lower Canada's official endorsement of the canal cost sharing bargain, he stated that the Canadians had already voted £10,000, and they would vote £25,000 plus £10,000 per annum for six years, according to Richmond "for the improvement of the navigation of the Ottawa River."[40] The two latter sums never were voted for the purpose stated by the Duke; nor is there evidence that the Lower Canadians were willing to contribute money to military canals further up the Ottawa than the immediate vicinity

---

36 For example, while there is considerable comment about communications improvements in the *Montreal Gazette* during this period, no mention is made about Canadian desires for Ottawa canals.

37 P.P. 1830-31 (135) IX, 15, "Canada Canal Communication," pp. 12-14, Report on military settlements, by Lieutenant Colonel Cockburn, Nov. 26, 1818.

38 W.O.44/19, p. 13, Treasury minute, May 25, 1819.

39 P.P. 1830-31 (135) IX, 15, "Canada Canal Communication," p. 16, Goulburn to Harrison, July 26, 1819.

40 C.O.42/182, pp. 143-44, Richmond to Bathurst, May 20, 1819.

of its confluence with the St. Lawrence.[41] Yet Richmond was suggesting that the Canadians were willing to divert at least a large portion of £85,000 into two military canals on the Ottawa estimated by Captain Mann to cost only £16,740. These figures make no sense and it is difficult to imagine how Richmond arrived at them honestly.

Nor did the Duke stop here. He also used Sherbrooke's misleading estimate of September 5, 1817, which, he said, amounted to about £50,000. He glossed over the fact that this sum applied only to the Lachine Canal. The Treasury had approved Sherbrooke's estimate and agreed to contribute half of the amount to the Lower Canadian project. Richmond therefore asserted that he felt himself authorized to appropriate £25,000 from Army Extraordinaries for the commencement of both the Lachine and Ottawa projects. He had in fact already done so. He had taken 300 shares in the Lachine Canal worth £15,000 and, since it seemed certain Lower Canada would provide £85,000, he intended to take the £10,000 left over from the British contribution of £25,000 in order to begin immediate work on the Ottawa. The latter statement of intention was actually a statement of another *fait accompli*. Under Richmond's orders Captain Mann and the Staff Corps were already spending the £10,000.[42] Whatever else these manoeuvres were, they were certainly irregular. Even if Richmond was only guilty of slips of the pen and an inability to add and subtract, he was still committing his government to the Ottawa project on false pretenses, on the pretext that the Canadians would share equally in meeting its costs. Furthermore, he proceeded without the prior sanction of the Treasury. Having thus made his contribution to the ultimate execution of the Ottawa-Rideau military waterway, he went on an inspection tour of the Rideau area, was bitten by his pet fox, and died of hydrophobia on August 10, 1819.[43]

Apparently no one in London saw reason to question Richmond's arguments or actions. Under the new governor-in-chief, Lord Dalhousie, the Staff Corps continued to work on the first part of the Ottawa project, the Grenville Canal.[44] It was almost two years later, on June 13, 1821, that Dalhousie wrote to Bathurst again about Richmond's arrangements. The governor mentioned in passing that the Lower Canadian Executive

---

41   The bill Richmond referred to was lost due to prorogation and never reintroduced. It was entitled a "Bill to improve the communication between the Provinces of Lower and Upper Canada by Internal Navigation and other purposes therein mentioned" and is located in R.G. 14, A-4, Vol. 2. It did contain a provision for Canadian canal construction on the Ottawa River, but only at the mouth of the river, not at Grenville, Carillon or Chûte à Blondeau where the military canals were later built. The preamble of the Lower Canadian Lachine Canal Act of 1821 (1, Geo. IV, Cap. VI, 1821) refers to the earlier Lachine Canal Act of 1819, but says nothing about provisions for Ottawa canals in either of these acts.

42   C.O.42/182, pp. 143-44, Richmond to Bathurst, May 20, 1819.

43   *D.N.B.*

44   R.G.8, C39, pp. 64-66, Mann to Bowles, Sept. 30, 1819. Mann reported to the governor that he had already spent £1,544 on the Grenville Canal. He was at this point turning over command of the work to Captain Henry du Vernet, Royal Staff Corps. See also ibid., p. 129, Du Vernet to Darling, Nov. 27, 1820, and R.G.8, C40, pp. 29-31, Progress report on the Grenville Canal, by Du Vernet, Dec. 18, 1820.

Council seemed to have doubts about the supposed Ottawa canal cost-sharing agreement, but stated that he intended to spend £8,000 more on the Grenville project during the summer out of funds already allocated by Richmond.[45] Bathurst replied very correctly that Dalhousie could keep spending on the Grenville Canal as long as Lower Canada met half the costs.[46]

Meanwhile the Lower Canadian legislature passed a new Lachine Canal act[47] which to some extent disentangled the situation perpetrated by Richmond and his predecessors. It took Dalhousie and the Colonial Office some time to figure out what the new financial position was. On August 2, 1822, A. W. Cockran, Dalhousie's secretary, attempted to clarify matters. He stated that by Bathurst's despatch of April 3, 1818, the imperial government had pledged to pay £35,000 for the Lachine Canal.[48] But in the new act of 1821 Lower Canada agreed to accept only £10,000. This left £25,000 at the British government's disposal for the Grenville Canal. Cochran then added that

> . . . though the authority of the Treasury of 25th May 1819 required that half the expense of this Canal should be borne by the Province, it appears to have been understood that the Province was exonerated from this charge, by having released, as it were, to Govt. £25,000 out of the £35,000 which under the dispatch of 3rd April 1818, it would have been entitled to claim from the Imperial Govt—which £25,000, be it remembered, exceeded what was originally estimated as the cost of the Grenville Canal. It was therefore thought to be fairly applicable to the entire completion of that.[49]

It seems that Cochran was reasoning backwards, justifying the imperial government's commitment to the Ottawa project by a Lower Canadian act which was passed after Richmond had made the commitment. In any case the Treasury had still not authorized any definite expenditure on Ottawa canals.

To get this authorization Dalhousie wrote to Bathurst on November 2, 1822, repeating Cochran's arguments and adding some of his own. He stated that he had just paid over to Lower Canada the £10,000 agreed on in the recent act, for which consideration all of His Majesty's vessels would receive free passage through the Lachine Canal. The remaining £25,000 he was applying to the Grenville Canal. These two canals were both essential, but Dalhousie stressed they were quite distinct projects; the Lachine Canal was civilian and the Grenville a purely military work. If this was clearly understood, he asserted, then all the original financial arrangements had been correct. Exactly how this could be assumed he did not explain. Having thus disposed of one problem, he moved to another.

---

45 C.O. 42/187, Dalhousie to Bathurst, June 13, 1821.
46 R.G.8, C39, Bathurst to Dalhousie, Aug. 8, 1821.
47 See note 40 preceding.
48 Richmond had calculated this sum at £25,000. How Cockran recalculated it is not revealed by the documents.
49 R.G.8, C40, pp. 104-08, Memorandum by Cochran, Aug. 2, 1822.

The Lower Canadians, Dalhousie stated, needed no more financial help for their civilian canal. But it seemed that the £25,000 which he had allocated to the Grenville Canal that very day had already been spent by the Staff Corps. Everyone knew, he argued, that the actual costs of large construction projects always tended to exceed original estimates because of unforeseen problems. It was therefore not surprising that the engineer in charge required another £25,000 to complete the Grenville Canal. Dalhousie strongly urged London to grant authority to continue expenditures on this vital military work. Just as Gother Mann had argued in 1800 about the St. Lawrence military canals,[50] the governor now expressed his confidence that eventually the Grenville Canal would pay for itself by tolls. For the present it was providing much needed work for destitute immigrants, was enabling them to settle the lands in its vicinity, and soon it would make all the lands between Montreal and Kingston, now "an immense Wilderness and Forest . . .", into populous settlements contributing to the prosperity of the Empire. Dalhousie added one more most telling point: "It might be presumptuous in me to urge the further evident Consideration that, to check the Work now, would be to throw away all the money already expended, with the addition of damages to Proprietors of Lands thro' which it passes: under all these explanations I entreat Yr Lordship's further interposition with H. Ms Government that I may be permitted to proceed on this great Public Work."[51]

Here were the classic arguments to solicit funds from the Treasury for military projects in the colonies, later to be repeated by Colonel John By and others. In particular there was Dalhousie's final point, which was almost an ultimatum. With it the manoeuvres of Sherbrooke, Richmond and Dalhousie reached culmination. These men had placed the Treasury in a trap from which it could escape only by writing off £25,000 as a dead loss. The military canal promoters had done more. They had passed the Grenville Canal off as a link between the two Canadian provinces, by implication as the whole Ottawa-Rideau waterway, when they knew perfectly well it was only the first small section, practically useless by itself. But once the Grenville work was completed it would be possible to show that unless the rest of the Ottawa-Rideau system was also built, the money already spent would have been thrown away on next to nothing.

The Colonial Office took over a year before it ventured to press Dalhousie's arguments on the Treasury. On November 10, 1823, Wilmot Horton replied to Dalhousie, enclosing a memorandum on "Water Communication between Upper Canada and Lower Canada" which repeated all of Dalhousie's own arguments.[52] A month later Horton sent this document to the Treasury.[53] The Lords Commissioners deliberated until April 9, 1824, and finally agreed to sanction Dalhousie's financial

---

50 R.G.8, C38, pp. 1-8, Mann to Hunter, Dec. 24, 1800.
51 W.O.44/32, pp. 118-21, Dalhousie to Bathurst, Nov. 28, 1822. This document also appears in C.O.42/191, pp. 302-05.
52 R.G.8, C40, pp. 220-23; Horton to Dalhousie, Nov. 10, 1823.
53 W.O.44/19, p. 14, Horton to Harrison, Dec. 10, 1823.

arrangements for the completion of the Grenville Canal.[54] During the intervening months the Staff Corps worked on the Ottawa River in the knowledge that further canals were needed at the Carillon rapids and the Chûte à Blondeau. These in turn would be useless for military purposes if the Rideau Canal was not also built. By 1825 the military canal promoters were ready to reveal these hard facts to the Treasury. Before turning to the Rideau project which from that year dominated Canadian defence schemes, however, the completion of the works on the Ottawa is worth noting.

After 1826 the Ottawa and Rideau projects proceeded simultaneously. The Staff Corps worked steadily from 1819 to 1829, under the unobtrusive authority of the Quartermaster General's Department and the immediate direction of local commanders-in-chief.[55] In February 1829 the Corps as such was disbanded, most of its personnel were absorbed by the Ordnance, and the completion of the whole military waterway became the responsibility of the latter department.[56] But the Ottawa and Rideau projects were kept separate, with separate accounts. Staff Corps officers and men continued their old duties under the new management and finished the Grenville, Carillon and Chûte à Blondeau canals comprising the Ottawa system on April 30, 1834.[57] While their costs exceeded estimates, in contrast to the men responsible for the Rideau Canal they did not become involved in major financial difficulties.

Captain J. W. Mann's original estimate came to only £16,740, although there is reason to believe he did not have the whole Ottawa system in mind.[58] The final cost was £312,009.[59] This discrepancy seems enormous, but Staff Corps personnel were not responsible for much of the increase in expenses. They began their canals on the Lachine Canal scale, with locks 108 feet long and 20 feet wide; then, in 1829 they were ordered to enlarge their locks to a length of 134 feet and a width of 33 feet as a result of Colonel John By's efforts to build as large a Rideau Canal as possible. They had to expand four out of seven locks on the Grenville Canal and all the locks on the other two canals.[60] That the Staff Corps personnel were careful with public money is indicated by the report of the House of Commons select committee of 1832, investigating Canadian military canal costs. The report censured Colonel By's extravagance with Rideau Canal funds but praised the officers working on the Ottawa project. While Rideau Canal costs always exceeded estimates by large amounts, in 1831 the Ottawa canal costs were £41,822 below the estimated figure. In referring to the latter canals, therefore, the committee

---

54 P.P. 1830-31 (135) IX, 15, "Canada Canal Communication," p. 18, Treasury Minute, Apr. 9, 1824.
55 W.O.44/32, p. 243, Mann to Byham, Feb. 27, 1828.
56 W.O.47/1456, p. 2025, Ordnance minute, Feb. 27, 1829.
57 C.O.42/253, p. 147, Butler to Hay, June 25, 1834.
58 See note 33 preceding.
59 W.O.44/15, p. 286, Treasury Minute, July 2, 1833.
60 P.P. 1830-31 (135) IX, 15, "Canada Canal Communication," pp. 67-68, Kempt to Murray, Feb. 12, 1829.

called "the attention of the House to the care and economy with which they appear to have been conducted. . . ."[61]

The completion of these canals, then, involved no difficulties for the government. Their interest lies rather in the way they were begun and used by those striving for a secure military route to Upper Canada, in particular by Drummond, Sherbrooke, Richmond and Dalhousie. Driven by their concern over the vulnerability of the St. Lawrence waterway which the Canadians wanted to improve and make even more vulnerable, they resorted to deviousness in order to circumvent the Treasury veto of 1816 and then extract sanction for the Rideau project. They succeeded because the government in London knew too little about the Canadians, geography, and colonial finances; it lacked the means to verify the assurances of its overseas agents. When Colonel By was appointed to build the Rideau Canal these defects still prevailed, and he exploited them even more blatantly. In his personal struggle to convert a barge canal into a seaway for ocean-going vessels he damaged both Canada's military security and his department.

---

61 P.P. 1831-32 (in 570) V, p. 4, Report of a select committee of the House of Commons on canal communication in Canada, June 29, 1832.

Chapter V

Launching the Rideau Project

By 1825 the military canal promoters were ready for their final step. Without the Rideau Canal the money spent on the Ottawa project would be wasted, a prospect which forced the government to yield to their demands. Their leader was now Wellington, adding to the effort the weight of his prestige, forceful personality and a reformed Ordnance Department ready for big overseas projects. As a result the Rideau plan was put into operation with undue haste. The financial arrangements were mishandled so that responsibility for canal expenditure was divided among several departments and individuals, and the Ordnance itself failed to make adequate preliminary cost and construction surveys. Most important the department selected Colonel By as superintending officer without adequate examination of his qualifications for so large a project; despite his long-standing reputation among admiring Canadians he proved to be a minor disaster to the British government, a prime example of the deficiencies in the imperial administration no less in the hasty way he was selected than in his subsequent performances.

The Rideau Canal campaign was actually resumed in 1824 by Sir Peregrine Maitland, the lieutenant governor of Upper Canada. At this time Dalhousie was still trying to get Treasury approval for the construction of the Grenville Canal. Maitland decided to renew agitation for the Rideau project because of the work of a committee of the Upper Canadian legislature appointed in 1821 under the chairmanship of John Macaulay to investigate ways of improving provincial water communications.[1] The commission sat for four years and among other things examined the feasibility of the Rideau route. Under its direction the civil engineer, Samuel Clowes, surveyed this route in 1823 and submitted an estimate for a canal with locks 100 feet long and 22 feet wide. He calculated the total cost at £69,783, a figure which, in the words of the Macaulay Commission's 1823 report, was "a sum absolutely insignificant when compared with the magnitude of the object, for the attaining which it would be applied."[2] Maitland

---

1 See Legget, *Rideau Waterway*, pp. 28-31, and Glazebrook, *History of Transportation in Canada* (1964 ed.), I, 78.
2 C.O.42/374, pp. 157-65, Report of a commission of the Upper Canadian legislature under the chairmanship of John Macaulay on internal navigation, Dec. 20, 1823.

forwarded this report to Bathurst in April 1824, stating that the province wanted British financial aid for this Rideau project. Because of the canal's great military and civilian importance, he urged that such support should be given.[3] Maitland's arguments, however, produced no effect, and the Upper Canadians themselves decided to abandon the Rideau route for the St. Lawrence.[4] But in 1825 Samuel Clowes prepared three more Rideau estimates, one of which was for a canal with 15 foot wide locks costing £145,000.[5] This latter figure did influence the commencement of the Rideau project. In its haste to start work the Ordnance resorted to Clowes' £145,000 sum as a basis for its own calculations of costs. The commencement of construction was speeded up as a result, with unfortunate consequences. London formed a misleading impression of the expense it was assuming, and Colonel By was able to blame Clowes for miscalculations he himself made.

Meanwhile Dalhousie undertook the final effort to gain official sanction for the military waterway. Writing from his home in Scotland on March 27, 1825, he pointed out to Bathurst that, because work on the Ottawa canals was now well advanced, it was time to start on the last link of the Ottawa-Rideau system. He reminded the colonial secretary that the imperial government had placed military settlers on the Rideau River "with a view to establish a safe communication during war by the retired line of the Ottawa. . . ." The population of these settlements was growing and a Rideau canal was becoming a pressing necessity. Since Upper Canada showed little inclination to contribute funds, the project would have to be purely military. Because of this, despite the fact that he had been urging the creation of a waterway previously, it might now be cheaper to think in terms of a road or railroad instead. In any case, Dalhousie concluded by asking that his letter should be given to the Duke of Wellington, who was sure to be interested and could at once order the Royal Engineers to investigate the best methods of proceeding.[6]

Dalhousie's request was timely. Wellington had advocated the construction of the Ottawa-Rideau communication in his Canadian defence memorandum of March 1, 1819, and already had firm ideas on the subject. His knowledge of Canadian geography and the military problems it presented was thorough, and he favoured the most advanced techniques, including the use of steam power, as means to improve communications between the provinces. He even proposed to extend the Rideau Canal to the west so that it would also form a direct route to Georgian Bay on Lake Huron.[7] The latter proposal was not taken up by the imperial government, but the Canadians later did adopt the Georgian Bay ship canal project which resulted in the construction of the

3   C.O.42/372, p. 158, Maitland to Bathurst, Apr. 1, 1824.

4   Glazebrook, *History of Transportation in Canada*, I, 78.

5   P.P. 1830-31 (135) IX, 15 "Canada Canal Communication," pp. 22-23, Report of a joint committee of the Upper Canadian legislature under the chairmanship of John Strachan and J. B. Robinson on water communications, Apr. 6, 1825.

6   C.O.42/204, pp. 108-09, Dalhousie to Bathurst, Mar. 27, 1825.

7   G.D.54/3/335, Wellington to Bathurst, Mar. 1, 1819.

Trent Valley Canal.[8] Besides his strong views about the Rideau project, by 1825 Wellington had also completed the reconstruction of the Ordnance and was ready to use it for large undertakings in Canada. On April 7, eleven days after Dalhousie wrote his letter, the Duke organized a commission of Royal Engineers under the chairmanship of Colonel Sir James Carmichael Smyth to investigate British North American defence needs in detail.[9] Four days later he ordered the commission to sail.[10] Smyth received instructions to follow the guidelines set out in Wellington's 1819 memorandum and to pay particular attention to the problems involved in completing the Ottawa-Rideau navigation as soon as possible.[11]

The commission, including Major Sir George Hoste, R.E. and Captain J. B. Harris, R.E. besides Smyth, apparently departed for North America on the same day its instructions were drawn up.[12] Five months later it presented a completed report to Wellington. In that time the commissioners crossed the Atlantic twice and examined British North American defences from the maritime provinces to the western end of Upper Canada.[13] The document they produced was the most detailed military analysis of the North American provinces to that time; it brought immediate action.

The commission, however, had done its work too hastily. Dalhousie complained that Smyth and his colleagues had not spent enough time on the scene to form sound opinions,[14] and this criticism was born out by subsequent events. The commission's figures for the cost of essential military works proved to be too low, although the minimum sum to secure the provinces from American invasion, £1,646,218,[15] was sufficiently large to horrify the Treasury. The estimate for the Rideau Canal was £169,000, leaving just under £1,500,000 for all other projects. On completion the canal alone cost in excess of £822,000,[16] representing a five-fold increase of the commission's estimate. The estimate for all other works rose by one-fifth from some £1,500,000 to £1,860,710 in 1828.[17] This discrepancy is largely explained by the fact that the Rideau Canal was actually finished—on an enlarged scale—while only a small propor-

---

8   Glazebrook, *History of Transportation in Canada*, I, 85.
9   C.O.42/205, p. 170, Wellington to Horton, Apr. 7, 1825.
10 Ibid., p. 174. Instructions to the Smyth commission, by the master general and board of Ordnance, Apr. 11, 1825. There is evidence that Dalhousie's letter of March 27, 1825, caused Wellington to send the Smyth commission to Canada. Wellington referred to this letter in writing to Smyth about the Canadian assignment, and Smyth in turn asked Dalhousie for advice before he left for North America. See ibid., p. 184, Wellington to Smyth, Apr. 12, 1825, and G.D.45/3/390, Smyth to Dalhousie, Apr. 8, 1825.
11 C.O.42/205, p. 174, Instructions to the Smyth commission by the master general and board of Ordnance, Apr. 11, 1825.
12 G.D.45/3/390, Smyth to Dalhousie, Apr. 8, 1825.
13 C.O.42/208, Report of the Smyth commission, Sept. 9, 1825.
14 G.D.45/3/390, Memorandum on the Smyth commission report, by Dalhousie, May 20, 1826.
15 Bourne, *Britain and the Balance of Power*, p. 39.
16 W.O.44/15, p. 297, Statement of the total cost of the Rideau Canal, by S. Thomas, Jan. 29, 1834.
17 Wellesley, *Despatches, Correspondence and Memoranda*, I, 81-82: Memorandum on the money needed for Canadian defence works, by C. G. Ellicombe, Mar. 1, 1828.

tion of the other projects was completed. Yet already in 1825 the sum of £1,500,000 was too large for the Treasury and the 1828 figure was quite unacceptable to the Lords Commissioners. Wellington and other Ordnance officers understood this problem, and therefore decided to carry out their plans by gradual stages, asking for small amounts on a per annum basis. Thus in 1826 they requested £50,000 from the Treasury for canals and another £50,000 for fortifications. Even so, the Lords Commissioners gave them only £25,000 for canals and nothing else.[18] While the Smyth commission recommendations were therefore too ambitious,[19] nevertheless a slow and careful construction policy might have resulted in the completion of most of the total scheme. But by 1834 over one million pounds were spent on the Ottawa-Rideau project alone, and the irate Lords Commissioners determined to prevent any large-scale expenditures for Canadian defence in the future. Hence the unexpected multiplication of Rideau costs did much to ruin the commission's plans.

Smyth and his colleagues were conscious of the need to persuade the Treasury that the initial Rideau estimate was reasonable and justified. All the same, they uncritically accepted Samuel Clowes' figure of £145,000 and merely added £24,000 to it for widening the locks to the 20 foot scale of the Lachine and Ottawa canals. For a total of £169,000 they proposed to build a modest barge canal designed to meet minimum military requirements. Because this figure was still very large, the commission approached the Canadians to see if they would change their minds and would contribute money, but they claimed they were too poor to help. To persuade the Treasury to accept the whole burden, therefore, the commissioners mustered the types of arguments which had already been used by previous military canal promoters. The government had already spent £60,000 on the Grenville Canal and another £50,000 would have to be provided to complete the other two Ottawa canals. Without the Rideau Canal all this money would be wasted. In the words of the commission's report, "the whole of this Water Communication would cost £279,000; but as it would be entirely in the hands of Government, the tolls would of course be collected on account of the Treasury; and in proportion to the rising prosperity and increasing commerce of the Province, the money advanced might be expected to be repaid. Excepting it is undertaken by His Majesty's Government, we are afraid it will never be executed."[20] So this absolutely indispensable project would ultimately cost the government nothing.

Wellington gave full support to the commissioners, forwarding their report to Bathurst on December 6, 1825, with a covering letter. He insisted that the government had to exert maximum effort to get the money for the whole defence scheme from Parliament immediately. It was inevitable that the Canadas would be lost to the empire if these funds were not granted. On the other hand, if the commission's report was fully implemented, the coming war in North America would not be very expensive. The

18 Bourne, *Britain and the Balance of Power*, pp. 40-41.
19 Ibid., p. 41, and Hitsman, *Safeguarding Canada*, p. 124.
20 W.O.44/19, p. 16. Report of the Smyth commission, Sept. 9, 1825.

Smyth plan was the cheapest way to defend Canada in the long run.[21] Wellington failed to get all he demanded despite these strong words, but the Rideau Canal project to which he gave chief priority was approved at once. After a series of verbal consultations between representatives of the Ordnance, Colonial Office and Treasury, on February 17, 1826, the imperial government agreed to execute the Rideau project.[22] The Ordnance immediately followed up the decision, and rushed into new errors compounding those of the Smyth report.

On March 10, 1826, General Gother Mann, the I.G.F., received orders from the office of the Ordnance to select two Royal Engineer supervisors for the major works on the Rideau and at Fort Henry in Kingston respectively.[23] Four days later Smyth wrote to advise Mann of the main points to be stressed in drafting official instructions for the canal supervisor, and referred to Lieutenant Colonel John By as the officer selected for this duty.[24] Thus By was chosen between March 10 and 14, so quickly that there was little opportunity to examine his qualifications. Furthermore, Mann at once rushed the Colonel off to Canada, failing to provide him with formal orders. When By arrived at Quebec on May 30, therefore, instead of orders—which would follow later—he carried only a selection of documents intended to guide his actions temporarily.[25] Among these was the Smyth report, a supplementary momorandum by Smyth, and a number of other papers and marginal notes supplied by the office of the Ordnance. In the memorandum Smyth again stressed the point that the Rideau locks had to be on the same scale as those of the Ottawa canals. He advised By to seek advice from the military engineers already working on Canadian canals, to consult the Upper Canadian Macaulay Commission report, and to read all printed information about the American Erie Canal. He suggested that civilian contractors be used to get the work done as quickly as possible, for an adequate military work force was not available.[26] But all this was informal advice, no substitute for direct orders; inevitably, the hastily chosen By would interpret these documents in his own way.

The office of the Ordnance confused matters even more when it explained to By the arrangements made for financing construction. The annual sums allocated to the project would be voted by Parliament as part of the Colonial Office budget. By was to carry his accounts as "Supplementary Ordnance Accounts." The Ordnance storekeeper in Canada, acting on the Colonel's instructions, would draw the necessary money from the Military Chest, controlled by the Commissariat, and the Chest would be reimbursed by the Colonial Office. To provide some central control over these complex arrangements

---

21 C.O.42/205, pp. 193-200, Wellington to Bathurst, Dec. 6, 1825.
22 W.O.44/24, pp. 569-70, Ordnance minute, Feb. 8, 1831. This minute is a précis of all transactions relating to the Rideau Canal between Apr. 1825 and Feb. 1831.
23 W.O.44/18, pp. 65-67, Griffin to Mann, Mar. 10, 1826.
24 Ibid., pp. 69-72, Smyth to Mann, Mar. 14, 1826.
25 *Montreal Gazette*, June 5, 1826, W.O. 44/15, p. 5, Ordnance minute, 1826, and W.O.44/18, pp. 144-47, Ordnance minute, May 27, 1826.
26 W.O.44/18, pp. 69-72, Smyth to Mann, Mar. 14, 1826.

the Ordnance—in London—would render an annual account of all Rideau financial transactions to the Colonial Office after getting this information from Canada, and the Colonial Office in turn would send the account to the Treasury. There was one additional point to complete the confusion. The office of the Ordnance stated that By was at liberty to start spending money without waiting to be informed how much Parliament would allocate to his project in the first year of construction. All he had to do was to consult Bathurst as to the sum he intended to spend each year. In short, By virtually had a free hand to use as much money as he wanted, presumably but not explicitly within the bounds of the Smyth Commission estimate of £169,000.[27] Here the Ordnance exceeded its authority; only the Treasury, with parliamentary consent, had the power to grant junior imperial officers such wide financial powers. With the division of responsibility among the Ordnance, Colonial Office, Treasury, the Colonel and the Canadian governor in chief, the likelihood of trouble was guaranteed. As events proved, By extracted maximum advantage from this situation; he selected from these loose instructions the points which gave him the fullest freedom and ignored the rest.

The Ordnance realized that these documents were inadequate. Richard Byham, the newly appointed secretary to the department, sent Bathurst all the Rideau papers for approval on April 3, 1826, when By was already in Canada. The colonial secretary was supposed to provide the money, so his approval was necessary before "proper instructions" could be drawn up for the Colonel.[28] Wilmot Horton replied on April 18 that Bathurst approved of everything and had nothing to add except that the canal should be built "with all possible dispatch; and his Lordship is of opinion, that it will be proper to authorize the contractor to commence as early in the season as circumstances will permit, without waiting for the passing of the annual grant."[29] Still further delays ensued, however, this time caused by Wellington. He was absent from the Ordnance on a foreign mission until the end of May,[30] and authorized the preparation of By's final orders only on June 15. In doing so he noted again that the Rideau project could commence without a parliamentary grant and commented on methods of construction. Then he issued two cautionary directives. All contractors used by the Ordnance should be kept firmly within the legal bounds of their contracts, and all officers in charge of the project had to be held within the normal chain of command. Colonel By in particular should under no circumstances be allowed to assume "any novel relation towards the authorities in Canada" despite the fact that he directed a project authorized and paid for by departments other than his own. It was essential to maintain normal channels of military communication in all proceedings even at the expense of delays and increased costs. If By wanted to make any changes in the existing canal plan he could do so only

27   W.O.44/19, p. 17, Smyth to Mann, Mar. 14, 1826, and W.O.44/16, pp. 24-25, Byham to the R.O. Quebec, 1826.
28   W.O.44/19, p. 16, Byham to Horton, Apr. 3, 1826.
29   W.O.44/18, p. 78, Horton to Griffin, Apr. 18, 1826.
30   Ibid., pp. 144-47, Ordnance minute, May 27, 1826.

after consultation with both the Ordnance and Colonial Office.[31] Wellington was here attempting to make sure that London authorities did not lose control of the Rideau project, but he was also too late.

On July 21, 1826, Colonel By's formal instructions were at last completed. Under the circumstances these orders turned out to be nothing but a compilation of the papers By had already taken with him and a few additional items.[32] Among them only Smyth's insistence that the Rideau locks had to conform to the Lachine Canal scale and Wellington's directives about strict conformity to military channels imposed restraints on the superintending officer. The main points were still those made by Bathurst. The canal should be built as quickly as possible and it was not necessary to wait for annual parliamentary grants. The upshot was that By's formal instructions were still too loose. Wellington, Smyth, Mann and Bathurst thus plunged into the Rideau project without taking adequate precautions to maintain control over the man in charge of construction once he was removed from them by the width of the Atlantic and the wilderness between Montreal and the site of the canal.

An officer with a keen sense for military propriety might have interpreted these instructions literally and sought the approval of higher authority for every step he took, as Wellington demanded. Strict conformity to orders might have minimized the effects of the mistakes thus far made. Colonel By, however, had strong opinions and ambitions; he interpreted his instructions according to his own designs. Historians who have written about him and the Rideau Canal appear to agree it was fortunate for Canada that the Colonel was this sort of man.[33]

Born in London on August 10, 1779, John By was the second son of a respectable middle-class civil servant, George By, employed in the London Customs House. Little is known of John's youth or early education but he was evidently intelligent and well-schooled enough to pass the entrance requirements of the Royal Military Academy at Woolwich and to obtain his commission as second lieutenant in the Royal Artillery at the age of twenty. The same year, 1799, he transferred to the Royal Engineers and in 1801 was promoted to first lieutenant.[34] After some routine service in England he went to Canada in 1802 and supervised the completion of one of the three small military canals on the St. Lawrence, at the Cascades. He was praised for the quality of his work and promoted to second captain in 1805.[35] From 1806 to 1811 he was stationed at Quebec primarily to assist in the extension of fortifications and showed initiative and technical skill by constructing a large-scale model of the citadel and town which was

31   C.O.42/379, pp. 131-35, Minute by Wellington, June 15, 1826.
32   W.O.44/18, p. 99, Ordnance minute, June 21, 1826.
33   See C. S. Blue, "John By: Founder of a Capital," *Canadian Magazine* XXXVIII (1912), 573-79; Hill, "The Construction of the Rideau Canal"; Hill, "Lieutenant Colonel John By"; and Legget, *Rideau Waterway.*
34   Legget, *Rideau Waterway*, p. 69, and Hill, "Lieutenant Colonel John By," p. 522.
35   See p. 35 above.

later sent to Woolwich to facilitate the planning of improved defences.[36] During 1811 he joined Wellington's army in the Peninsula and took part in the siege of Bajados. The next year he became superintendent of the Royal Gunpowder Mills at Faversham, Purfleet, and Waltham Abbey, and in 1814 he constructed the small arms factory at Enfield Lock; he designed a new type of truss bridge with an extremely wide span and was subsequently promoted to brevet major. In 1821 he was relieved of active duty because of peacetime economies in the armed forces but was nevertheless promoted to lieutenant colonel three years later.[37] Thus, when he arrived in Canada in 1826 to take up a very welcome new post, By was an experienced engineer officer of 47, married, with two young daughters; his career had been creditable though not spectacular, his rate of promotion about average for the corps, and his technical competence proven. The Rideau project, however, was his first large independent command.

Concerning his appearance and character at this time there is little first-hand information aside from scattered references in official documents and the papers of a few contemporaries. Aparently he was a stout, florid, dark-haired man of medium height and upright bearing; he was energetic, hard-working, meticulous, and even-tempered. The majority of those he worked with seem to have both liked and respected him. More important, however, he was also a visionary aspring to speed up Canadian economic prosperity and to enhance the importance of the provinces within the empire. He was intensely interested in the aspirations of Canadians. His efforts to develop the pioneer settlement of Bytown into a well-ordered community gave him the title of founder of Ottawa, the nation's capital.[38] Canada's parliament buildings stand on lands he set aside for grand public purposes and there are reminders of his influence in many other places, down to the Bytown Inn. Similarly his imprint is evident along the whole route of the Rideau Canal, which is regarded as a great contribution to Canadian material progress. The Colonel has earned much credit for building this work on a grander, more enduring scale than his superiors demanded. Regrettably By was not given more scope to build the even greater canal he called for, a canal which would have formed a seaway from the Atlantic to the Great Lakes. If By had been supported rather than censured by the imperial government Canada would have had this seaway four generations before it was completed, and Montreal and Toronto instead of New York and Chicago might have today dominated the commerce of North America.[39] So run the speculations of those who see the Colonel as a far-seeing and benign builder of Canada, surrounding the man with a myth which takes many forms. For example, one writer has been moved to poetry:[40]

36   W.O.55/859, p. 430, R.H. Bruyeres to General Morse, Nov. 16, 1810, and p. 432, By to Morse, Feb. 7, 1811.
37   Legget, *Rideau Waterway*, p. 70, and Hill, "Lieutenant Colonel John By," p. 523.
38   See Blue, "John By: Founder of a Capital."
39   See Legget, *Rideau Waterway*, pp. 68-83.
40   William Pittman Lett, quoted in M. A. Friel, "The Rideau Canal and the Founder of Bytown," *Women's Canadian Historical Society of Ottawa, Transactions*, I (1901), 31-35.

As o'er the past my vision runs,
Gazing on Bytown's elder sons,
The portly Colonel I behold,
As plainly as in the days of old,
Conjured before me at this hour,
By memory's undying power;
Seated on his great black steed,
Of stately form and noble breed,
A man who knew not how to flinch,
A British soldier every inch,
Courteous alike to low and high,
A gentleman was Col. By.

Another writer asks

Is it stretching imagination too far to conjecture that the selection of him [By] as
the man to build the Rideau Canal was made by the great Duke of Wellington
himself? His service in the Peninsular War would probably have brought him
to the attention of the Duke. He could not have done what he did on the
Rideau without having displayed similar ability at earlier tasks. The combination of any such demonstrated ability and his nine years of engineering
experience in Canada would have made him an ideal candidate.[41]

In fact it was Gother Mann who selected By—in haste—and Wellington disapproved of
the choice.[42] Yet imagination and conjecture have thus far prevailed and nourished the
myth of the man who, from the Canadian viewpoint, made lasting contributions to the
development of present-day eastern Ontario.

For the imperial government and the Ordnance, however, By proved to be an
unfortunate choice. Throughout the nineteenth century the agents of empire serving
abroad from time to time forced their superiors in London to alter fixed policies and
assume uncomfortable responsibilities. Control over such men was inadequate, and
Colonel By benefitted from this in a manner particularly regrettable to the London
authorities.[43] For them his activities were damaging, unjustifiably draining the public

---

41 Legget, *Rideau Waterway*, p. 70.
42 By was selected by General Mann, probably because the General remembered By's work on the St.
    Lawrence canals from 1802 to 1806. During that period Mann had been By's commanding officer in
    Canada. See W.O.44/18, pp. 65-67, Griffin to Mann, Mar. 10, 1826, and ibid., pp. 69-72, Smyth
    to Mann, Mar. 14, 1826. When Wellington became aware of By's attitudes early in 1827 he
    expressed serious doubts about his fitness. See W.O.44/32, p. 215, Memorandum by Wellington,
    Jan. 27, 1827.
43 In terms of Galbraith's "man on the spot" thesis, as expressed in his "The 'Turbulent Frontier' as a
    Factor in British Expansion" and *Reluctant Empire*, Colonel By is one of the worst types of "men on
    the spot" in nineteenth century imperial history. Galbraith referred to important proconsuls such as
    Thomas Stamford Raffles in Singapore between 1819 and 1824, Sir Andrew Clarke and Sir William
    Jervois in Malaya between 1873 and 1876, Sir Benjamin d'Urban at the Cape between 1834 and
    1836, and Sir George Grey in South Africa between 1854 and 1861. These men expanded the
    imperial domain in defiance of London policy. On a lower level there are also examples such as John
    Graves Simcoe in Upper Canada between 1792 and 1796, a visionary determined to build the

purse and detrimental to Canada's military security. Because of the resultant determination to reduce defence expenditures in colonies, perhaps the Canadians also lost something from military contributions to civilian development in the long run.

From May 30, 1826, when he arrived in Canada, By took until early February 1827 to complete a preliminary survey of the Rideau route.[44] In the meantime, however, he already began to express strong views about his assignment. On July 13, 1826, well before leaving Montreal for the Rideau, he wrote as follows to General Mann:

> I have the honor to report than on examining the Military defences of Canada, it appears selfevident that, by forming a steamboat navigation from the River St. Lawrence to the Lakes, would at once deprive the Americans of the means of attacking Canada; and would make Great Britain mistress of the trade of that vast population on the borders of the Lakes, of which the Americans have lately so much boasted, and to secure the trade have expended immense sums of money in cutting Canals, which Canals would in the event of our Steam boat navigation being completed, ultimately serve as so many outlets for British manufactored [sic] goods. I therefore feel it my duty to observe, that all the Canals at present projected are on too confined a scale for the increasing trade of Canada and for Military service they ought to be constructed of sufficient size to pass the Steam boat best adapted for navigating the Lakes and rivers of America, which Boats measure from 110 to 130 feet in length, and from 40 to 50 feet in width, drawing 8 ft. water when loaded, and are capable of being turned to military purposes without any expense, as each boat could carry four 12 Pounders and 700 men with great ease. It is therefore evident, that the moment our canals and Locks are completed on this scale, we shall not only possess the trade of all that immense population on the borders of the Lakes, but also have Military possession of the Lakes; for by having the power of collecting our forces at any one point with a rapidity of motion, that no land movement can equal, the Lakes must remain in our possession; and consequently Canada rendered perfectly secure from attack.[45]

By therefore urged that the Rideau, Welland and Grenville canals should be built large enough to pass steam boats 130 feet long and 50 feet wide, and that the inadequate Lachine Canal should be bypassed by a series of locks on the northern side of Montreal Island. Furthermore, the Richelieu River should be improved in the same way in order to attract American commerce northward to Montreal. He was confident these projects could be completed in four to five years. Then, by adding a "trifling work" at the Falls of

---

greatest British colony of all who was largely unsuccessful. (See S. R. Mealing, "The Enthusiasms of John Graves Simcoe," *C.H.A.R.* [1958], 50-62, W. R. Riddell, *The Life of John Graves Simcoe* [Toronto, 1926], and S. F. Wise, "The Indian Diplomacy of John Graves Simcoe," *C.H.A.R.* [1953], 36-44.) In a sense By was similar to Simcoe in his aims and achievements. He was also unique in so far as other noted "men on the spot" added territory to the imperial possessions while he extracted huge sums of money from an unwilling Treasury and Parliament.

44 Douglas Brymner, ed., *Report on Canadian Archives, 1890* (Ottawa, 1890), p. 79, By to Mann, Feb. 6, 1827.

45 W.O.44/19, pp. 199-204, By to Mann, July 13, 1826.

St. Mary between Lake Huron and Lake Superior, Montreal would have an uninterrupted water communication all the way to the western end of Lake Superior.

The imperial government should execute this whole project itself because no part of it could be safely entrusted to the Canadians. By pointed out that despite all precautions the Lachine Canal had cost £107,000, twice as much as expected. "I mention this to prove that the Civil Engineers in this country, are either very ignorant, or have designedly Estimated these Works at less than they can be executed for." By's steam boat canals, more carefully planned, would not be expensive. The Welland and Rideau canals would cost £400,000 each, the Grenville £100,000 and the Richelieu £150,000. While the total figure was £1,200,000, tolls and other economic benefits would soon repay this expenditure. He concluded by requesting orders to proceed with this project, adding that "I am therefore of opinion whether the Works I have now the honor of projecting cost one, or two Million, it is of no moment, when compared with the positive economical means they will afford of rendering Canada perfectly secure. . . ." Thus one or two million pounds were sums of "no moment" to a man who would spend five times the original estimate on the Rideau Canal, a man who castigated Canadian civil engineers for spending twice the Lachine Canal estimates. Simultaneously he was projecting his own huge estimates without having seen the ground, much less surveyed his various proposed canal routes. The Colonel was venturing beyond the bounds of a Royal Engineer's military and professional duty as well as misconstruing the specific nature of his assignment.

Before this letter reached London his superiors made preparations to arrange for the financing of the Rideau Canal. The actions they took indicated that they did not in fact consider that By had a free hand to spend money when and how he wished. In a minute dated August 10, 1826, Wellington asked his subordinates at the Ordnance to decide on the 1827 sum for Canadian canals so that he could submit this figure to cabinet on September 5.[46] Smyth completed the calculations on August 17. For 1826 Parliament had given £15,000 for Canadian water communications, of which £10,000 was allocated to the Ottawa canals and £5,000 to By. This left £164,000 still to be voted for the Rideau Canal, which would probably take five years to complete. Therefore £32,800 would be needed per annum, and Parliament should be asked to supply this sum for the financial year 1827. Smyth added a word of caution: "I take the liberty respectfully to remark, that whatever may be the amount it may be determined upon to apply for, it is of the utmost consequence that Lieut. Colonel By should have the most early notice. . . ."[47] By's instructions said nothing of parliamentary votes before starting to spend. They did, however, stipulate that the total cost of the Rideau Canal would be £169,000. As long as the Colonel managed to stay within the limits of this estimate it would not matter too much if in any given year he exceeded the sum voted by Parliament.

---

46 Brymner, *Report, 1890,* p. 72, Minute by Wellington, Aug. 10, 1826.
47 W.O.44/19, p. 18, Smyth to Mann, Aug. 17, 1826.

But he soon made it clear to the Ordnance that he did not wish to be bound by the £169,000 figure. His letter of July 13 arrived in London shortly after August 17. In passing it on to Smyth for comment General Mann reacted mildly, saying that "although the Lt. Colonel's zeal is praiseworthy yet I did not fall in with his opinion either of the practicability or advantage of what he proposes. . . ."[48] Smyth expressed stronger opinions. He dismissed outright the idea of canals fifty feet wide; a width of twenty feet was adequate for all military purposes including the movement of gunboats. If the government undertook to build canals for large warships, there would be no limit to the size of the locks needed. For civilian traffic, from which the hoped-for toll revenues were expected, twenty-foot canals would also suffice. Smyth felt that By could not have examined the country properly or he would have realized that his larger canals would cost much more than he expected: "The impression upon my mind is that he would encounter greater difficulties than he seems to be aware of." The Colonel's proposal to improve the Richelieu River was especially upsetting to Smyth who felt that the river was already too open to American invaders: "It does not appear to me that Lt. Col. By has taken a judicious view of the Military Features of defences of Canada in proposing to improve the navigation of the River. . . . If he could add to the impediment, it would, in my opinion, be more advantageous to His Majesty's Service."[49]

Wellington's reaction to By's letter was similar to that of Smyth. Lord Fitzroy Somerset, in his secretarial capacity, relayed the Duke's views to Mann on September 1. The master general regarded the idea of larger canals as entirely impractical, and the commanding Royal Engineer in Canada was to tell By he was not to contemplate improvements to the navigation of the Richelieu River under any circumstances. By was to "proceed with activity to exercise the Service upon which he is employed, without altering any part of the Plan proposed. . . ."[50] As usual Wellington's restraining directive arrived in Canada too late to divert By from his chosen path. On October 1, 1826, he was again urging Mann to approve the seaway project.[51]

In the meantime the effects of the haste in which the Rideau project had been launched were creating problems elsewhere as well. Colonel Elias Durnford, the commanding Royal Engineer in Canada, received a copy of By's official instructions only in September 1826. Durnford asked the governor in chief to issue all necessary orders and make all arrangements to enable By to commence construction; apparently the commanding Royal Engineer did not know that By's preliminary instructions had at least stipulated that he was to start work as soon as he arrived in Canada. It was especially important for the governor to obtain legal title to the lands needed for the canal.[52] At the same time Durnford ordered By to begin surveys at both ends of the proposed Rideau

---

48 Brymner, *Report, 1890*, p. 75, Mann to Fitzroy Somerset, Aug. 26, 1826.
49 Ibid., pp. 73-75, Smyth to Mann, Aug. 23, 1826.
50 W.O.44/24, pp. 649-50, Fitzroy Somerset to Mann, Sept. 1, 1826.
51 W.O.44/18, p. 9, By to Mann, Oct. 1, 1826.
52 P.P.1830-31(135)IX, 15, "Canada Canal Communication," p. 41, Durnford to Darling, Oct. 14, 1826.

route.[53] On October 25 Governor Dalhousie wrote that he had received no orders concerning the project from Bathurst. Nevertheless, on the strength of By's instructions from the Ordnance he informed Durnford that he was prepared to authorize the commencement of the canal.[54] It was already becoming obvious that overlapping authority was creating confusion.

The biggest problem emerging from this confusion concerned the Rideau lands. On December 22 Smyth wanted to know why the Colonial Office had not yet obtained the necessary lands it had promised. Government had to act quickly or the proprietors would inflate their prices as soon as they found out about the canal. Furthermore, it was unjust to expect that Parliament should have to pay for these properties. Upper Canada would benefit greatly from the canal, and therefore should be made to contribute the site. Smyth urged the Colonial Office to extract a commitment from the Upper Canadian legislature as soon as possible.[55] Mann repeated Smyth's complaints and exhortations on January 1, 1827, noting that Wilmot Horton of the Colonial Office had assured the Ordnance that all steps to get the land had been taken back in April 1826.[56] In fact no action was taken in Canada until the end of January 1827,[57] and nothing was accomplished until the Upper Canadians passed their Rideau Canal Act on February 17. The Act met none of the demands of Smyth and Mann; instead it proved a source of trouble to the Ordnance.[58]

Colonel By also complicated the start of construction by creating delays. In November 1826 he still had not submitted any progress report to the office of the Ordnance.[59] His first report was dated February 6, 1827, and stated that he had completed his survey of the whole Rideau route and had begun construction with an initial work force of two companies of Royal Sappers and Miners.[60]

Early in December 1826, while still busy with preliminary surveys, By had received the reactions of his superiors to his large-scale canal proposal. Despite the rebukes of Wellington and Smyth he refused to abandon his plan and assured Mann that he was "confident the Rideau Canal will be completed agreeably to my Instructions on the Scale of the La Chine & Grenville Canals in four Years, although I have great doubts whether it can be performed for £169,000, but I will not venture to give a decided opinion on this Subject until I have well examined the whole line."[61] But he expressed his concern about the fact that his superiors could not see the great value of his large

---

53 Ibid., p. 42, Durnford to By, Oct. 14, 1826.
54 Ibid., Darling to Durnford, Oct. 25, 1826.
55 W.O.44/19, pp. 275-78, Smyth to Mann, Dec. 26, 1826.
56 Ibid., p. 19, Mann to Byham, Jan. 1, 1827.
57 P.P. 1830-31 (135) IX, 15, "Canada Canal Communication," p. 42, Horton to Byham, Jan. 31, 1827.
58 *Statutes of Upper Canada*, 8 Geo. IV, Cap:1, A.D. 1827.
59 C.O.42/379, p. 159, Byham to Horton, Nov. 27, 1826.
60 Brymner, *Report, 1890*, pp. 78-81, By to Mann, Feb. 6, 1827.
61 W.O.44/19, pp. 203-09, By to Mann, Dec. 6, 1826.

canal idea: "I regret extremely that Major General Sir J. C. Smith [sic] should for a moment Conceive, that I would have undertaken the construction of the Rideau Canal had I felt as wanting in practical knowledge as I imagine he conceives I must be from his remarks. . . ." The Colonel then introduced a series of technical arguments to prove that fifty-foot canals were far more practical and useful than twenty-foot canals. For only £50,000 more he could easily build them on the large scale. The question which he left unanswered was £50,000 more than what? After beginning with a warning that £169,000 was too little for a small canal, and admitting that he had not yet done enough surveying to provide a more realistic figure, he still insisted that he could enlarge the project two-and-one-half times for an additional £50,000.

When Wellington saw this second letter he became seriously worried. "I must say that I am not at all at my ease about the employment of Lt. Colonel By in the execution of this great Work, after the perusal of his letters—I have never seen upon any subject such a performance as the last; in which he proposes so many alterations, without stating for what object. He must adhere strictly to his Orders respecting the Dimensions of the Locks."[62] After the usual delay the Ordnance sent By Wellington's categorical order to follow the original plan.[63]

By received the order early in July 1827 but made no formal acknowledgement. Instead he now sought to overcome the immovable resistance of his superiors by applying indirect pressure on them. On July 7 he appealed to Sir Peregrine Maitland, lieutenant governor of Upper Canada, complaining about the master general's restriction on the size of the Rideau Canal and asking for help with the promotion of his enlarged canal plan.[64] Shortly afterwards John Carey, the editor of the Upper Canadian York *Observer*, wrote to Prime Minister George Canning, asking him to sanction a Rideau Canal large enough to take steam vessels. Carey himself would contract to build it for only fifteen per cent more than the estimated cost of the small canal.[65] By also resorted to the *Montreal Gazette*, the mouthpiece of Montreal merchant interests. The newspaper took up his cause and referred to him as "our friend on the Rideau."[66] And well it might, because he proposed to give the Canadian merchants the seaway to the interior lakes they so much desired, at the entire expense of the British Exchequer.

Wellington got wind of By's campaign. On October 11, 1827, he wrote about certain suspicious events in Canada to Lord Anglesey, who had succeeded him as master general.

> I enclose a printed Paper which has been circulated in the King's Dominions in America, which I have reason to believe was written by Lt. Colonel By of the Engineers. If you will refer to the papers on the Subject of the Inland

---

62 W.O.44/32, p. 215, Memorandum by Wellington, Jan. 27, 1827.
63 W.O.46/29, pp. 38-39, Minute by Fitzroy Somerset, Mar. 13, 1827.
64 C.O.42/381, pp. 388-89, By to Maitland, July 7, 1827.
65 C.O.42/382, p. 149, Carey to Canning, July 10, 1827.
66 *Montreal Gazette*, Aug. 16, 1827.

Navigation in Canada, you will see that that officer proposed that the Canal to be constructed between the Ottawa & Kingston should have its locks of the dimensions of 180 feet by 60, instead of 108 feet by 20, with a view to the Navigation of the Canal by Steam Vessels; which proposition I overruled. The consequence of the renewal of the discussion in this form, is to render the people of Canada, and possibly those of the Country and even Parliament itself, dissatisfied with this work. . . . [The] Expence of the Rideau Canal Calculated at £169,000 would have amounted to about £500,000 if the large dimensions had been adopted; and if the large dimensions had been added as they must have been to the La Chine Canal, the Grenville Canal, the Chûte à Blondeau & others upon the Ottawa to render the System compleat the Expense would not have been less than a Million Sterling; and I believe it will be admitted, that no Minister could have gone to Parliament to propose the grant of such a sum.[67]

In the light of later events Wellington's estimates for the large canal system were too low, but they were far more realistic than anything suggested by Colonel By.

The Colonel was unaware of the concern he was causing in London. On December 10, 1827, he again wrote to Smyth in his usual fashion. He stated that Commodore Robert Barrie, R.N., commanding naval officer in Upper Canada, strongly supported the larger project, and now that "the whole line of Canal is laid open & everyone can see that nature has formed the Canal & that the large Locks can be completed in three years at the trifling increase of £50,000 to my Estimate, all persons are anxious for the large Locks. . . ."[68] At the same time By was doing little to complete the first detailed cost estimates for the small Rideau Canal. After a full year in Canada he still had no precise figures. On May 21, 1827, the House of Commons had to make a formal request to the Colonial Office for Canadian canal estimates.[69] The next day Wilmot Horton provided the calculations, which were then printed by order of the House. They were still based on Smyth's total of £169,000 with an additional allowance for contingencies; the sum was to be divided into four annual grants of £41,000.[70] All London authorities based their calculations on these figures until December 1827, when Colonel By revised them drastically.

In the meantime he carried on his work on the assumption that he did not have to concern himself with parliamentary grants. While his superiors assumed he was staying within the limits of the £5,000 granted to him for 1826 he was moving to widen the area of his personal spending powers. He asked the Ordnance for authority to enter into building contracts with civilians entirely on his own discretion, without recourse to normal Ordnance regulations; the reason he gave was his anomalous position among the

---

67   C.O.42/219, pp. 192-97, Wellington to Anglesey, Oct. 11, 1827. Anglesey replaced Wellington as master general on April 30, 1827. See *D.N.B.*

68   W.O.44/32, pp. 157-59, By to Smyth, Dec. 10, 1827.

69   C.O.42/213, p. 6, House of Commons address to His Majesty, May 21, 1826.

70   P.P. 1826-27 (161) (380) XV, 285, 283, "Water Communication in Canada," Estimates by Wilmot Horton, May 22, 1827.

overlapping authorities of several departments of state.[71] Wellington had specifically ruled that By was not to have "any novel relation towards the authorities in Canada,"[72] yet here the colonel claimed just such a "novel relation." The Ordnance approved his request, adding a rather fatuous proviso that he could only depart from the normal regulations governing contracts when it was absolutely necessary.[73]

The Rideau Canal Act of February 17, 1827, also extended his freedom of action. It did not give the imperial government free and clear title to all the lands needed for the Rideau project as the Ordnance had hoped. The government could indeed take all the lands it needed, but only by paying a price set by a jury of the local inhabitants. The agent specified by the act as authorized to buy land was Colonel By, who could make purchases in his own name.[74] To the authority his instructions gave him to proceed without waiting for parliamentary grants, therefore, By added the power to make contracts and buy lands as he saw fit. While most authorities in London continued to assume he was an Ordnance officer building a twenty foot wide canal at the rate of £5,000 in 1826 and £41,000 per annum during four subsequent years, the Colonel proceeded at the rate of expenditure he himself thought necessary.

Such were the consequences of hasty organization. Wellington rushed Smyth and his commission to Canada, the commission hurried its report, the Ordnance selected the canal builder and drafted his instructions too quickly, and Colonel By plunged into grand seaway proposals without pausing for preliminary surveys. Wellington and Smyth tried to regain close control too late. The Rideau finances were already so confused it was nearly impossible to fix responsibility for expenditures on any one department or individual; Colonel By was already armed with independent powers sufficient to escape the restraints of his superiors. Until December 1827 only Wellington and Smyth saw that the Rideau project was getting out of hand. Thereafter By moved to commit the government to a canal plan much closer to his own designs, and made it plain that he had the strength to coerce even the House of Commons.

---

71 W.O.55/864, p. 53, Byham to the R.O. Quebec, June 13, 1827.
72 C.O.42/379, pp. 131-35, Minute by Wellington, June 15, 1826.
73 W.O.55/864, p. 53, Byham to the R.O. Quebec, June 13, 1827.
74 *Statutes of Upper Canada*, 8 Geo. IV, Cap: 1, A.D. 1827, W.O.44/42, pp. 300-05, Memorandum by Smyth, Dec. 3, 1827, and W.O.44/28, pp. 38-42, Memorandum by Ordnance solicitor James Smith, Feb. 1836.

Chapter VI

The Expansion of the Rideau Project

Between December 1827 and June 1828 Colonel By extracted official permission to build a canal larger than the London authorities wanted. Although the work was not commensurate with his ambitions, its extension represented a personal victory over Wellington and Smyth—a victory made more notable because it was achieved in the teeth of opposition to By's initial estimate for the canal on the Smyth commission scale. Besides good fortune, the Colonel achieved success by manipulating data with much skill, but in a way which revealed many defects in the machinery of colonial control. Ultimately Whitehall recognized the faults demonstrated by his tactics.

In December 1827, when By disclosed the fact that he would need close to £500,000 instead of £169,000 for the small scale canal, his superiors expressed consternation. A search for the culprits showed only that adequate information on the Rideau project was lacking; indeed, a committee of Royal Engineers, working partly from By's own despatches, found little to report. Wellington was nevertheless convinced the Colonel was guilty of misconduct and insisted on a second investigating committee to check By's work in Canada. Before this committee left for the Rideau the retiring master general, Anglesey, decided By was right about everything. Huskisson, the colonial secretary, accepted Anglesey's judgement. When the second committee sailed, therefore, it was not instructed to investigate By's conduct as Wellington had intended, but instead carried orders to approve the construction of large canals if these seemed feasible. As a result the imperial government unknowingly assumed a commitment to an eventual expenditure of over one million pounds.

By began construction in the summer of 1827 but did not send his cost calculations until October 4, eighteen months after arriving in Canada. He also sent the figures in an indirect fashion. Writing to Gother Mann, the I.G.F., he stated that an officer under his command, First Lieutenant H. Pooley, R.E., was on his way to London to deliver all the necessary information in person. Pooley would also argue the case for large canals, and By expressed confidence that his superiors would be persuaded, among other reasons because "the whole country was regretting that this beautiful water communication should be blocked up with small Locks. . . ." The Colonel added that work on the first

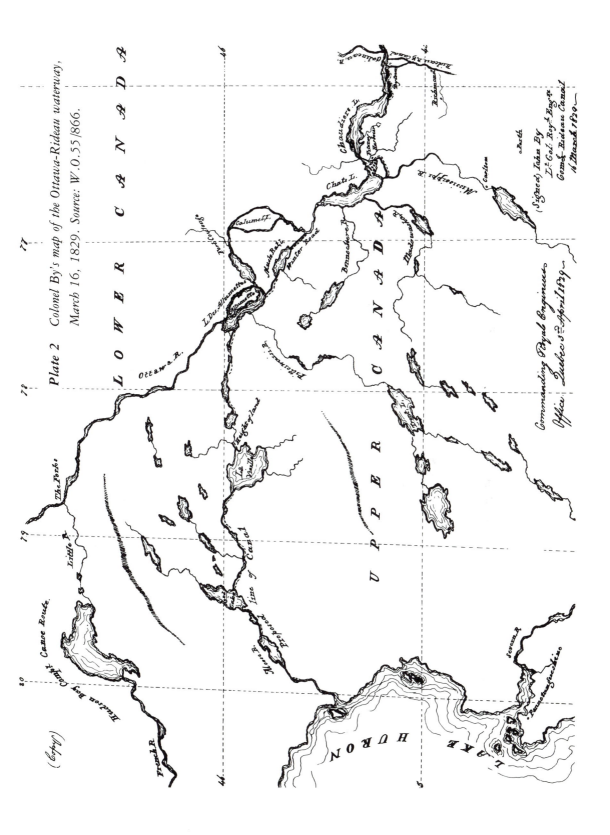

Plate 2   Colonel By's map of the Ottawa–Rideau waterway,
March 16, 1829. Source: W.0.55/866.

three locks was going well and that two Lachine Canal commissioners had been amazed to find his masonry costing less than a third the amount they had paid for theirs. In conclusion By stated he was making faster progress than expected. At a rate of £100,000 per annum for the next three years he would finish the whole canal by August 12, 1830.[1]

Although By wrote that he was sending Pooley on October 4, the covering letter Pooley took with him was dated November 1, 1827. Further delays had ensued. This second letter was again addressed to Mann and was substantially a repetition of the October 4 letter. The Colonel, however, did assert that he had begun the first three Rideau locks as ordered, on the small scale. He also explained that Dalhousie had suggested sending Pooley as the most certain method of delivering the Rideau documents in time to present them to Parliament for approval.[2]

This explanation for employing Pooley indicated that By knew Parliament was waiting for detailed figures and would scrutinize his calculations carefully. The critical document containing the estimate was a scantily itemized list of costs covering little more than one page; the total came to £474,000.[3] In his October 4 letter he had asked for £100,000 per annum for the next three years to finish the project. Therefore he must already have spent about £174,844 in 1826 and 1827. Parliament, it will be recalled, had provided only £46,000 for those two years. Thus By seems to have exceeded the total of the original estimate even before notifying his superiors of the extent of the necessary upward revisions.

The reaction to his figures revealed the gulf between his conception of his powers and that of higher authorities involved in the Rideau project. In Canada Dalhousie received the revised estimate before it reached London. While he seems to have been one of By's supporters,[4] he refused to contemplate the request for £100,000 per annum, insisting that By could not exceed £41,000, the sum voted by Parliament. Only the Ordnance or Colonial Office could authorize an increase.[5] In London the reaction was much sharper. By's figures arrived just before December 10, 1827. Huskisson, the colonial secretary, was displeased but restrained himself in official correspondence. He ordered E. G. Stanley to write to Richard Byham requesting that the Ordnance supply complete information on all Rideau expenditure and investigate the possibility of any further cost increases. The Colonial Office asked that Ordnance officers responsible for canal construction be made to exercise the strictest economy and to keep meticulous accounts showing where every penny went. In Stanley's words, Huskisson was "extremely unwilling" to make further applications for money to Parliament.[6]

---

1   W.O.44/19, pp. 197-98, By to Mann, Oct. 4, 1827.
2   P.P. 1830-31 (135) IX, 15, "Canada Canal Communication," p. 43. By to Mann, Nov. 1, 1827.
3   Ibid., pp. 43-34, Appendix to By to Mann, Nov. 1, 1827.
4   Brymner, *Report, 1890*, p. 81, By to Dalhousie, Oct. 26, 1827. In this letter By thanks Dalhousie for praising his work.
5   Ibid., p. 84, Darling to Durnford, Nov. 29, 1827.
6   W.O.44/19, pp. 82-83, Stanley to Byham, Dec. 10, 1827.

Anglesey, the new master general, was much less restrained in his reaction. Bedridden with a severe illness when he received By's figures, he penned two excited private notes to Huskisson. He wrote that "I am extremely sorry to inform you that the Report lately received from Lt. Colonel By of the Royal Engineers is most distressing. The Estimates for the Rideau Canal brought forward by the late Board of Ordnance & sanctioned by Parliament turn out to be grossly erroneous."[7] After a brief account of the Rideau project, starting with the Smyth commission and ending with By's £474,844, he continued:

> Lt. Col. By has sent home Lt. Pooley an intelligent officer of Engineers for the purpose of giving information and explanation. He states that the Civil Engineer then employed purposely & avowedly misled the Colonial as well as the British Government. This may be, but it appears to me wonderful that a work of the Magnitude & Importance of that of the Rideau Canal, should have been commenced without a careful examination of the Estimates by the Engineers of the Department. There must be much negligence or imprudence somewhere you will see the necessity of an early and careful investigation.

On the strength of Pooley's word Anglesey accepted that there had been no misappropriation of public funds but added he was totally opposed to By's enlargement proposal. In the second note to Huskisson he expressed similar dismay.

> Your discretion will prevent you from publicly noticing my observations upon the negligence or the imprudence of beginning a great work without an Estimate. It is instead so unlike the usual course of proceeding of my predecessor [Wellington] that I cannot but think there must be some mistake upon the subject.
>
> That the Rideau Canal *must* be finished upon the scale originally approved by the Duke of Wellington, I have not the slightest doubts, but there are objections to its enlargement. First the expence would be greatly increased—2dly Its completion would be postponed—3dly There is a doubt if there would be a sufficient Command of Water at all times for the higher levels. I imagine that it will be impossible to find the Means of finishing the Canal now on hand within the time at first calculated upon, in consequence of the immense increase of the Estimate & we must be content to do that in 6 or 8 Years which it was proposed to affect in 4.[8]

Anglesey had discovered the basic cause for the discrepancy in cost calculations, decrying "the negligence or the imprudence of beginning a great work without an Estimate." Whether the master general intended it or not, this amounted to stating that every authority connected with the Rideau project thus far shared the blame. He therefore asked Huskisson to avoid discussing his sweeping criticisms publicly. It seems he rejected Pooley's argument that Samuel Clowes, the Canadian civil engineer, had "purposely & avowedly" misled everyone. Such an explanation fixed the blame on an

7   C.O.42/213, pp. 235-36, Anglesey to Huskisson, Dec. 12, 1827.
8   Ibid., pp. 239-40, Anglesey to Huskisson, Dec. 13, 1827.

outsider at one level but put the burden of guilt on the Smyth commission at another. The commission had accepted Clowes' figure as sound.

While others in London continued to express concern over By's estimate, Anglesey soon moderated his rigidly critical attitude. He ordered the assembly of all Rideau documents, examined them in detail, and on December 18 reversed his initial opinions. Among the documents was a summary of the whole canal problem written by the inspector general of fortifications. Gother Mann argued that the original estimate was basically the work of a civil engineer who had neglected to consider military needs. Mann confessed that it was probably impossible to produce a full explanation of all other causes which might have contributed to the rise in cost calculations. But this was because the "only information in the inspector general's office which may be considered as fully detailed is Colonel By's estimate," and some supplementary details from By and Pooley. Mann was confident that By had acted properly in all respects; it was Clowes who was guilty of misleading the British government. Furthermore, Mann added that according to the documents By had so far spent only £32,622 but had entered into "agreements"—contracts—involving undisclosed additional sums. Therefore the Colonel's full financial commitments were unknown, but the I.G.F. was sure they were perfectly acceptable. There would be no further increases in estimates.[9]

Thus Mann adopted By's arguments, and Anglesey in turn adopted Mann's. On December 18 he wrote Huskisson stating that "there seems to be no means of escaping from the expence. The great question appears now to be within what time Government will direct that the Work shall be completed." The master general felt it would be best to give By £100,000 per annum in order to keep to the old schedule. Perhaps it would also be wise to adopt the large canal plan, especially since the Colonel estimated the cost of enlargement to be very reasonable.[10] Anglesey here was practically paraphrasing By; he was now among the Colonel's converts. As a precaution, however, on January 4, 1828, he appointed a committee of Royal Engineers composed of Major General Sir Alexander Bryce (president), Colonel John T. Jones and Lieutenant Colonel Edward Fanshawe to double check the Rideau plans and estimates.[11]

In the meantime, on December 27, 1827, Wellington took time from his political activities to write another memorandum on the Rideau problem. Unlike Anglesey he held to his earlier opinions. He commended the decision to convene an investigating committee as "most judicious," and expressed the view that By appeared to have "lost sight entirely of the Plan and Estimate" of the Smyth commission despite repeated reminders to obey orders. The Duke admitted that the original Rideau estimate might have been unsound, but By had not proved, by his own or Pooley's arguments, that the upward revision in costs was justified. Therefore an additional committee of engineers

---

9 Ibid., pp. 245-49, Ordnance minute, by Mann, Dec. 17, 1827.

10 Ibid., pp. 243-44, Anglesey to Huskisson, Dec. 18, 1827.

11 P.P. 1830-31 (135) IX, 15, "Canada Canal Communication," pp. 44-45, Bryce to Mann, Jan. 23, 1828.

should be appointed to investigate the Rideau project in Canada. Until both committees submitted reports, Wellington urged that no more than £40,000 could be allowed for the canal in 1828. Nor could By be permitted to enter into any contracts without prior approval of both the Ordnance and the Colonial Office.[12]

In this document, then, Wellington implied that By had been insubordinate on two counts. First, he had disobeyed orders by changing the original Smyth commission plan. Second, even if it proved that the commission's plan had been faulty, By was still guilty of negligence; he had failed to inform his superiors that the original plan and estimate were unsound until December 1827. Concerning the first count, it is possible Wellington thought By had actually begun his canal on a large scale. The Colonel's demands for larger canals were the only detailed messages he had been sending to London before Pooley arrived. In July 1826 By had stated that a Rideau canal on the *large scale* would cost £400,000,[13] a sum coinciding more closely with his first detailed estimate than with the Smyth commission figure. In December 1826 he had expressed doubts that the original sum of £169,000 would be enough for a small scale canal but had given no indication of how much would be needed.[14] On October 4, 1827, he had requested £300,000 more to complete the project, which either meant that he was throwing out figures at random or that he had already spent £174,844—more than the Smyth commission sum.[15]

A close reading of the documents the Colonel sent to London creates the impression that at best he was careless with his calculations; at worst he was deliberately evasive, perhaps because he began construction to provide for large scale locks as soon as these were sanctioned. Wellington read By's paper in this light but Anglesey did not.

The Colonial Office took up the Duke's recommendations immediately. Horton, writing to Stanley on January 1, 1828, expressed his decided approval, especially of Wellington's suggestion that By should have his powers to enter into contracts restricted. He added that it would be a good idea to try to transfer the Rideau Canal to the colonial authorities as soon as it was finished in order to escape the burden of maintenance costs. The Colonial Office also agreed to restrict By to an annual expenditure of no more than £40,000.[16] General Mann at once wrote to Lieutenant Colonel W. M. Gosset in Canada instructing him to inform By of these developments. The rate of Rideau expenditures had to be slowed as much as possible.[17] While this warning was crossing the Atlantic and the two investigating committees were preparing to start work, on February 25, 1828, the Colonial Office informed the Ordnance that Huskisson intended "to submit to Parliament an Estimate for the sum of £41,000, on account of the Rideau

12   C.O.42/219, pp. 182-89, Memorandum by Wellington, Dec. 29, 1827.
13   W.O.44/19, pp. 199-204, By to Mann, July 13, 1826.
14   Ibid., pp. 203-09, By to Mann, Dec. 6, 1826.
15   Ibid., pp. 197-98, By to Mann, Oct. 4, 1827.
16   C.O.42/219, pp. 180-81, Horton to Stanley, Jan. 1, 1828.
17   W.O.44/15, p. 9, Mann to Gosset, Jan. 4, 1828.

Canal, without pledging Government as to the amount to be required for this service another year. . . ."[18]

This information as well as Mann's warning to By came too late. At the end of January William Carr, Viscount Beresford, replaced Anglesey at the Ordnance. The change in command produced delays, but by March 17 Beresford knew his job well enough to point out to Huskisson that the decision to limit the Rideau expenditures had been futile. According to the latest report from Canada

> . . . you will see, that Lieut. Colonel By could not have been made acquainted with the intended limitation of expenditure for this year in sufficient time to prevent his making the contracts he therein communicates to have been completed; those contracts having been made, and the contractors in all probability having made their arrangements, and even commenced operations, we can only be liberated from those engagements (I speak of those made 1st February last) by very onerous concessions, which would be a dead loss. . . . In short, it is obvious that Lieut. Colonel By has laid down work for this year that will take about a third of the sum he estimated for the whole, or about £140,000 (taking into consideration what has been already expended), as the contractors go to complete the whole in three years from January last.[19]

By had written to the Ordnance that he had been proceeding on the assumption he was expected to press on as fast as possible, and already had committed the British government to contracts far in excess of £41,000 some three weeks before Mann's restraining order arrived.[20] In short, if his superiors refused to approve his actions they would not only lose the chance of finishing the canal on schedule but would also have to throw away large sums in compensation to contractors.

Beresford's announcement of this new problem came between the report of the Bryce committee and the departure of Wellington's second committee for Canada. The Bryce committee report was completed in eighteen days and was almost a total vindication of By. The Colonel had not deviated from the original plan except in minor details justified on technical grounds. As for his estimate, the committee considered it "on the whole to have been formed with much care and accuracy." The only change Bryce and his two colleagues recommended was a reduction in the thickness of masonry amounting to a saving of £9,000. They accepted By's contention that the Canadian civil engineer Samuel Clowes was at fault for misleading the British government, stating that

> On the whole we are of opinion that after allowing Mr. Clowes much credit for skill and industry in exploring and marking the best general line for effecting this Water Communication, under many difficulties occasioned by the state of the country, and with probably little professional assistance, his estimate for

18   W.O.44/19, p. 23, R. W. Hay to Byham, Feb. 25, 1828.
19   W.O.44/33, pp. 239-40, Beresford to Huskisson, Mar. 17, 1828.
20   W.O.55/865, p. 187, By to Mann, Mar. 5, 1828, and C.O.42/216, p. 93, Dalhousie to Huskisson, Mar. 28, 1828.

executing the necessary work is quite inadequate, and with his report, are [sic] rather calculated to show the practicability of the measure, than to give an accurate calculation of the expense of effecting it. This conclusion, we think, will appear well founded, on a careful consideration of the numerous plans, prepared with great labour and attention by Lieut. Colonel By, and his officers, and which were all necessary before any satisfactory design or estimate for the Canal could have been made.[21]

Turning to By's fifty-foot lock proposal, the committee noted that the Colonel still considered he could widen his canal for an additional £50,000 plus an extra £3,000 for widening the locks already under construction. The members of the committee agreed this bigger canal would have great but not "any immediate advantages," and accepted the figure of £53,000 without question. They concluded by agreeing with By's estimate as it stood and by declaring he had followed his orders to the letter. They did not think "that detailed estimates founded on accurate measurements and levels could be prepared and transmitted until the second summer." By could not have submitted his estimate any earlier.

Smyth had meanwhile become a member of the committee and added a short dissenting opinion. He repeated his earlier conviction that the enlargement of the Rideau Canal would necessitate an expensive enlargement of the Ottawa canals as well. However, his warning had little effect, and the Bryce committee which worked entirely from By's information fully concurred with all of that officer's actions. By's calculations proved proportionately a more severe underestimate of the final cost than the estimate of scapegoat Clowes was in relation to By's figure of £474,844.

The Bryce committee report confirmed Anglesey's decision to support By. On January 26, three days before he left the Ordnance, he wrote to Huskisson stating that the idea of enlarging the canal should not be abandoned. "On the contrary, I think it is a subject which ought to be fully discussed and considered in all its bearings, not only with reference to the defence of the country, but as connected with its trade and revenue."[22] Huskisson was persuaded. On March 14 he wrote Anglesey's successor, Beresford, that he disagreed with Smyth about the enlargement proposal. There would be advantages in enlarging the Rideau locks even without enlarging those on the Ottawa. If the committee which investigated By's work in Canada agreed that the Rideau locks could be widened for only £53,000 extra, Huskisson felt it would be "advisable to leave it to their discretion to authorize Lieut. Colonel By to proceed with the construction of the locks either to the increased dimensions of fifty feet, or of any intermediate size between twenty and fifty. . . ."[23] Even the shock he received three days later when Beresford told him By had already committed the government to an

---

21 P.P. 1830-31 (135) IX, 15, "Canada Canal Communication," pp. 45-50, Report of the Bryce committee, Jan. 22, 1828.

22 W.O.44/19, p. 20, Anglesey to Huskisson, Jan. 26, 1828.

23 W.O.44/32, pp. 241-42, Huskisson to Beresford, Mar. 14, 1828.

expenditure of £140,000 failed to move him from his position.[24] While regretting By's unauthorized entry into contracts, the colonial secretary still considered the Colonel's proceedings justified by local circumstances and remained willing to sanction the large canal scheme as soon as the committee going to Canada approved it. Therefore Huskisson asked Beresford to stress the desirability of large locks in the instructions given to the second committee of engineers. In fact he already assumed the revised plan would be adopted and now considered By's effective estimate to be £474,844 plus £53,000, a total of £527,844.[25]

Thus, from the negative reactions of Wellington, Smyth and initially Anglesey, Huskisson moved to a position of full support for all of By's plans. Because of Huskisson's latest opinions the instructions for what became the Kempt committee bore little relation to those proposed by Wellington. The Duke had hoped to have this committee at work by the end of March and had intended it to investigate By's conduct as an officer under specific orders.[26] Probably because of delays caused by the change in masters general it took the Ordnance until March 27 to draw up the committee's instructions. These stated that if By had acted with proper regard for economy and for the interests of the colony, he was to be authorized to spend £105,000 in 1828, which was in addition to £61,000 the Colonel supposedly had spent already. More important, the Ordnance emphasized that if the Rideau locks could in fact be enlarged for By's estimates the committee was empowered to order enlargement. Among the documents included in the committee's instructions were the Bryce committee report, Clowes' plans and estimates and By's plans and estimates.[27] The Smyth commission papers were not included, nor were any of the critical memoranda written by Wellington and Smyth. In short, these instructions were so favourable to By he could have drafted them himself. His persuasiveness, good fortune and particularly his luck that Wellington was no longer master general, had served the Colonel well.

While the Kempt committee was still being organized Huskisson apparently decided it would approve all of By's proposals. It was instructed to authorize the expenditure of no more than £105,000 in 1828, yet on April 9 Huskisson officially applied to the Treasury for £120,000 plus a supplementary amount of £79,000 for that year.[28] The Lords Commissioners granted the application.[29]

The colonial secretary's expectations were justified. The Kempt committee completed its assignment within three months of receiving its instructions. According to the Bryce committee at least two summers were needed for the officers on the Rideau to

---

24   W.O.44/33, pp. 239-40, Beresford to Huskisson, Mar. 17, 1828.
25   P.P. 1830-31 (135) IX, 15, "Canada Canal Communication," pp. 51-53, Huskisson to Beresford, Mar. 26, 1828.
26   C.O.42/219, pp. 182-89, Memorandum by Wellington, Dec. 29, 1827.
27   P.P. 1830-31 (135) IX, 15, "Canada Canal Communication," p. 53, Instructions of the Kempt committee, Mar. 27, 1828.
28   Ibid., p. 54, Hay to G. R. Dawson, Apr. 9, 1828.
29   Ibid., p. 54, Treasury minute, June 27, 1828.

do enough work to produce an accurate estimate.[30] Yet the Kempt committee spent well under two months on the site. Sir James Kempt wrote to Huskisson explaining that he did not arrive at Montreal from Nova Scotia to take charge of the committee until after June 12, and that he was in a hurry to return to his province.[31] He and his colleagues, Lieutenant Colonel Fanshawe and Lieutenant Colonel G. G. Lewis, R.E., also took time to examine the Ottawa canals and to mark sites for depôts and blockhouses along the Rideau route. The haste with which the committee worked is reflected in the rather superficial report it produced sixteen days after June 12.

The committee members approved of By's actions. They felt that "he has, in accordance with what he believed to be the spirit of his instructions, pushed forward the work, and excited a degree of exertion throughout the department, which few individuals could have accomplished." As for the problem caused by the contracts, the members felt they could only sanction an accomplished fact:

> Contracts are entered into for four-fifths of the navigation from the Ottawa to Kingston; and such preparations made for the progress of the work, that, although the amount of expenditure will greatly exceed that contemplated in the Instructions, the Committee had *no alternative*, but either to suspend their sanction for the further advance, and thereby involve Government in a certain loss for detention and breach of contract, or to authorize Lieut. Colonel By to proceed. . . .[32]

Had the Colonel already exceeded the figure of £105,000 which the committee was allowed to authorize? The report does not make clear if he had spent more than £41,000 or the newly sanctioned £105,000, but the latter sum is implied. Moving to the problem of enlargement, the committee rejected a fifty foot width but ruled that the canal should be widened from twenty to thirty-three feet. By was authorized to proceed on this scale to avoid sacrificing "any large portion of the expense already incurred or engaged for by specific contracts. . . ."

Were the committee members here referring to the fact that By had begun work on the assumption he would eventually build fifty foot locks? They did note that the Colonel already had done some extra digging which would ease the building of larger locks. By had explained that he hoped to open the canal for timber rafts as soon as possible. Lieutenant Colonel Fanshawe, writting of the problems involved in deepening the enlarged canal, stated that "even to obtain the five feet [depth], part of the Rideau Lake and Kingston Waters will require a considerable sum; but the service had been contemplated in Lieut. Colonel By's original Estimate."[33] This Kempt committee evidence is not conclusive but it does suggest By may well have been disobeying orders which restricted him to the Smyth commission scale.

---

30 Ibid., pp. 45-50, Report of the Bryce committee, Jan. 22, 1828.
31 Ibid., p. 55, Kempt to Huskisson, June 28, 1828.
32 Ibid., pp. 55-59, Report of the Kempt committee, June 28, 1828.
33 Ibid., p. 61, Fanshawe to Mann, June 30, 1828.

Concerning By's cost calculations, apparently the confident prediction about a fifty foot canal costing no more than £53,000 extra was now obsolete. The committee members pointed out that it was insufficient just to widen the locks; they also had to be deepened, and the Ottawa and Lachine canals would have to be enlarged as well. Such a project would cost an additional £250,000. By himself now stated that a fifty foot Rideau Canal would cost £599,176, not merely £474,844 plus £53,000. It seems he had previously neglected to add in increased labour costs and more expenditure for land and compensation to land owners. The thirty-three foot canal, he asserted, would cost £576,758. The committee reduced this figure to £558,000 and expressed confidence that this would be the final and accurate estimate for the redesigned canal.[34]

The upshot was that after making firm assurances he could build a fifty foot canal for only £53,000 extra By was now asking £101,913 extra for a thirty-three foot canal and the committee saw nothing unusual in this increase. The members were in a hurry and kindly disposed toward a fellow officer with a difficult job in the wilderness. They did in fact uncover indications that By's calculations were unreliable and that he might have begun work in defiance of orders to build a small canal. These were matters which Wellington had originally wanted them to examine, but such a task was not required by their instructions. Consequently, they allowed By to carry on and also removed doubts about his capabilities.

In short, the imperial government lacked means for close and continuous surveillance of officers working in remote regions of the colonies like the Rideau area. By's superiors were neither able to gather enough information about his activities nor adequately process what information they did have to conclude that he was the wrong man to execute so important a project. Anglesey and Huskisson, for example, appear to have decided that only By had sufficient local knowledge and therefore had to be trusted. For the Canadians that trust was fortunate because it held long-term benefits for Ontario, but for London it was misplaced. By's continued activities entangled his government and the Ordnance in mounting difficulties. His expenditures continued to exceed successive predictions of the final costs, and there was no way to stop his overspending.

---

34   Ibid., pp. 56-59, Report of the Kempt committee, June 28, 1828.

Chapter VII

Completion of the Canal and Financial Chaos

Colonel By finished the Rideau Canal between June 1828 and May 1832 with money which he extracted from the British taxpayer's pocket despite the House of Commons. His superiors repeatedly had to pay bills incurred without their approval or knowledge. It was hard to ascertain which department controlled any given budgetary item because responsibility for colonial military expenditures was too divided. Also, each department used its own disbursement and accounting procedures, which were not understood by others; furthermore, there was no centralized audit system to keep a check on colonial defence expenses. Colonel By gained advantage from all these defects.

In 1828 the Treasury and House of Commons were not fully aware of this situation. However, By soon demonstrated that a minor Ordnance officer stationed in a remote possession could determine the sums his government spent overseas. In doing so he forced the guardians of the public purse into a belated effort to tighten their control over colonial finances. In the course of this attempt the Treasury, unable to prevent the outlay of over a million pounds, turned against the Colonel, the Ordnance, and colonial defence projects in general. All plans to complete the whole Ottawa-Rideau navigation on By's enlarged scale were stopped; thus, for most military purposes the waterway remained twenty feet wide. The Ordnance could get no funds to remedy this wasteful defect, and very little money for all the other essential works recommended by Wellington. The finished Rideau Canal itself was impressive in size and appearance, but Colonel By executed it in a way which left the Ordnance with more problems than benefits.

The Kempt committee report reached London in October 1828 and, after some consultation among departments, was accepted by the Colonial Office.[1] On November 22, R. W. Hay informed the Treasury that the final Rideau estimate, at last a reliable figure, was £558,000. Of this By already had spent £166,000, and therefore needed

---

1 W.O.44/19, p. 29, Hay to Byham, Oct. 22, 1828.

£392,000 in three annual instalments to finish his project by 1831. Hay also mentioned that £80,000 had been spent on the Ottawa canals, but because of the necessity to redesign these to coincide with the new size of the Rideau Canal, the total estimate for them was now £176,640.[2] The Treasury, apparently confident that assurances from the Ordnance and Colonial office were still worthy of trust, accepted the new figures without comment. The Lords Commissioners agreed to request £130,067 for the Rideau Canal and £32,233 for the Grenville Canal from Parliament in 1829.[3]

But Colonel By was already changing the estimates, and on November 20, 1828, he sent a progress report to the inspector general of fortifications. He claimed his work had been slowed because of sickness among the workers and because he had been limited to an annual expenditure of only £41,000 prior to the Kempt committee report.[4] This was clearly devious, since he had earlier admitted that he had laid down work far in excess of £41,000 per annum before June 1828.[5] He said, further, that he was now spending £158,317, and wanted £137,216 per annum for the next three years to finish his project by August 12, 1831.[6] By's estimate was now several thousand pounds larger than the Kempt committee figure of £558,000.

His report arrived at the Ordnance early in January 1829 and was forwarded to the Colonial Office with a strong recommendation that By should be granted the annual sum he wanted.[7] The Colonial Office received the report on January 16 but made no comment.[8] The Colonel had also sent an identical report to Sir James Kempt, now governor-in-chief of Canada.[9] Kempt, however, did not forward this report to the Colonial Office until April 1.[10] When this duplicate arrived at the Ordnance via the Colonial Office, Richard Byham again urged that By's request for £137,200 should be approved. Because it was now too late in the year to ask Parliament for this sum, Byham suggested that By should be given £140,000 per annum in 1830 and 1831.[11] The Colonial Office passed Byham's request to the Treasury on July 28.[12]

On previous occasions the Treasury had accepted such requests without complaint. This time, however, the Lords Commissioners expressed concern. They pointed out that the Kempt committee figure of £558,000 had been accepted as final by all departments involved. According to their latest calculations, By now needed £18,000 more, and

2   P.P. 1830-31 (135) IX, 15, "Canada Canal Communication," p. 64, Hay to Stewart, Nov. 22, 1828.
3   Ibid., p. 65, Treasury minute, Dec. 23, 1828.
4   Ibid., pp. 65-66, By to Mann, Nov. 20, 1828.
5   Ibid., pp. 45-50, Report of the Bryce committee, Jan. 22, 1828, and pp. 56-59, Report of the Kempt committee, June 28, 1828.
6   Ibid., pp. 65-66, By to Mann, Nov. 20, 1828.
7   Ibid., p. 65, Byham to Hay, Jan. 12, 1829.
8   Ibid., p. 66, Hay to Byham, Jan. 16, 1829.
9   Ibid., pp. 85-86, By to Couper, Nov. 20, 1828.
10 Ibid., p. 81, Kempt to Murray, Apr. 1, 1829.
11 Ibid., p. 80, Byham to Hay, July 8, 1829.
12 Ibid., p. 80, Hay to Stewart, July 28, 1829.

they wanted an explanation; they wanted to see all the relevant documents. Until that time they refused to ask Parliament for more money.[13]

The Ordnance replied by revealing a most curious circumstance. It seemed "that the Lieutenant-Colonel does not appear to have been in possession of the reduced estimate for the Rideau Canal formed by the Committee of which Sir James Kempt was President."[14] Therefore the Ordnance recommended that By should be supplied with a copy of the calculations on the basis of the Colonel's own figures. The committee had authorized him to build a thirty-three foot wide canal at a specific rate of annual expenditure. It is little wonder, then, that the Lords Commissioners were showing signs of distrust for Ordnance assertions.[15]

The Colonel's reply to a request for explanation failed to reassure the Treasury, although at face value his progress report of December 30, 1829, contained ample justification for all his actions. He began by stating that "I have always reported the sum estimated for the Rideau Canal as the probable, not the positive sum; and it was, and still is utterly impossible to state the exact amount that will be required to complete that service. . . ." Yet, to the contrary, it will be recalled that the Colonel was always positive about nominating figures. The £50,000 needed to widen his canal is a case in point. By went on to claim that when Smyth had informed him in London that he was to build the Rideau Canal for £169,000, he had verbally "remonstrated against the smallness of the sum. . . ." By insisted that he had predicted that his canal would cost £400,000 as early as August 13, 1826, in a letter to Mann. After making detailed surveys he had calculated the cost at £474,899, but this, he emphasized, was the minimum. For example, he had not included "the expenses of the Civil and Military Establishments necessary to carry on such extensive works; wishing to show the actual amount of work indispensably necessary to form the proposed Water Communication. . . ."[16] Therefore By charged that he had not altered his own estimates—his letter to Mann of August 13, 1826, was proof of this. He had informed his government he would need at least £400,000, and his detailed estimate was quite close to this predicted sum. If By's superiors had been unpleasantly surprised by the first official figure of £474,844, they themselves were at fault for ignoring his warning of 1826.

This argument does not stand up under close scrutiny. Ordnance files contain no record of a letter from By to Mann dated August 13, 1826; instead, there is one dated precisely one month earlier, the letter which so upset both Wellington and Smyth.[17] In it the Colonel did indeed mention that the Rideau Canal would cost £400,000, but he was writing about a canal with locks fifty feet wide rather than twenty. He was predicting that he could build a canal two-and-one-half times larger than the one he was ordered to build for no more than £400,000. Even if he had written to Mann again

13   Ibid., pp. 80-81, Treasury minute, Aug. 4, 1829.
14   Ibid., p. 86, Butler to Stewart, Aug. 28, 1829.
15   Ibid., pp. 87-88, Treasury minute, Sept. 10, 1829.
16   Ibid., pp. 110-13, By to Durnford, Dec. 30, 1829.
17   W.O.44/19, pp. 199-204, By to Mann, July 13, 1826.

exactly a month later stating that the small-scale canal would also cost the same sum, his assertions of December 30, 1829, still make little sense. The surviving documents fail to show any indication that By warned London about the need to revise cost calculations for the small canal until the arrival of Pooley in England in December 1827.

The Ordnance, however, did not question the Colonel's assertions. It may be that the staff at the office of the Ordnance searched for the supposed letter of August 13, failed to find it—probably because it was never written—and assumed it contained what By told them. Alternatively they knew By's facts were twisted but decided it was more politic to accept his word. To question him openly was to impugn the proceedings of the entire department.

Having explained away his 1827 estimate By continued his report by stating that the Kempt committee had induced him to revise the 1827 sum upward to £576,758. But the committee had not informed him of its own lower estimate of £558,000. Therefore the blame for confusing calculations in London lay with Kempt and his colleagues. They had also limited him to an expenditure of £105,000 for 1828. His precise complaint in this respect is interesting:

> . . . no one step has been taken to enable me to comply with the instructions, as all contracts formed by the Commissariat Department in February 1828, which embrace nearly all the works on the line of the Canal, still remain in force. . . . I have no controul [sic] over the expenditures, the contracts being so worded that the contractors can demand payment as their works progress; and the fatal effect of the lake fever in the summer and autumn of 1828 (which still continues, though in a less alarming degree) has so increased their expenses that they are all exerting themselves to complete their works next season; and this is the cause of my disbursements for the present year amounting to £211,354/7/6¼ instead of £130,666/13/4 as authorized, notwithstanding my efforts to retard the expenditure as much as possible; and this gives me every reason to suppose that £200,000 will be required for 1830. This is on the supposition that the estimate is ample, and that no failure will take place in any part of the works, which, in such extensive waterworks, is almost improbable; but from the great success I have hitherto met with in those works, I trust all will succeed and that I shall have the honour of opening the Steam-Boat Navigation from the Ottawa to Kingston on the 12th of August, 1831.[18]

By concluded confidently that there would be no more upward revisions of the total cost.

Again he had spent more money than was authorized. The Treasury had reluctantly given him £130,666 for 1829; he had spent £211,354. Even more reluctantly the Treasury had given him £140,000 for 1830, but By wanted £200,000. This unhappy situation, he stated, was not of his making. In ordering him to confine himself to authorized sums the Kempt committee had asked the impossible. The Commissariat department had entered into contracts forcing the upward revisions.

---

18 P.P. 1830-31 (135) IX, 15, "Canada Canal Communication," pp. 110-13, By to Durnford, Dec. 30, 1829.

According to By, therefore, the responsibility for confusion over Rideau Canal expenditures rested with the Kempt committee and, more directly, with the Commissariat. The latter had drawn up contracts binding By as well as the government. Opposed to such claims, however, it seems unlikely that the Commissariat would have entered into contracts without instructions from the superintending engineer. Indeed, By's attempt to shift the blame becomes even more questionable when it is remembered that in June 1827 he received authority to make contracts on his own initiative.[19] He did instruct the Commissariat to draw up contracts for him; it seems incredible that department was forcing unwanted contracts on By.

The day after writing this report the Colonel prepared a briefer one for Mann. He sent copies of both to Sir James Kempt, who forwarded them to the Colonial Office on February 12, 1830; the Colonial Office sent them to the Treasury on June 14, 1830. By this time the Ordnance must have received its copies, but instead of taking them to the Treasury, the Treasury had to take the Colonial Office copies to the Ordnance on June 18. The Lords Commissioners wanted the master general's opinion of By's revised figures as well as additional information. On June 23 Sir Alexander Bryce, who had replaced Mann as inspector general, briefly replied that the Ordnance regarded By's last estimate of £576,757 as accurate and necessary.[20] Yet the Ordnance already knew that the Colonel had again raised the figures. Another report from By had arrived just after June 15. On August 17 the office of the Ordnance informed the Treasury that both By and the Kempt committee had left out an additional £166,686 worth of works from their estimates. The final cost of the canal was now calculated at £693,448, which meant that £256,777 would be needed in 1831 to finish the project on schedule. Even these sums might prove to be too small.[21]

In Canada By himself and Colonel Durnford thought that the extra expense of blockhouses and delays caused by accidents and illness would now put the final cost up to £762,679.[22] In an attempt to explain to the governor-in-chief why By had exceeded the latest parliamentary grant by £54,000 Durnford stated that the increases were the fault of no-one, especially not of By; he was doing an excellent job under adverse circumstances which made rises in costs virtually inevitable. In any case the accelerating expenditure could only be halted at the risk of lawsuits for breach of contract.[23]

---

19 W.O.55/864, p. 53, Byham to the R.O. Quebec, June 13, 1827.
20 P.P. 1830-31 (135) IX, 15, "Canada Canal Communication," pp. 95-96, By to Mann, Dec. 31, 1829; p. 94, Kempt to Murray, Feb. 12, 1830; p. 94, Hay to Stewart, June 14, 1830; p. 108, Treasury minute June 18, 1830; p. 110, Byham to Stewart, June 23, 1830; p. 113, Minute by Bryce, June 15, 1830. Note that Bryce replaced Mann as inspector general of fortifications early in 1830.
21 Ibid., pp. 114-16, Butler to Stewart, Aug. 17, 1830.
22 Ibid., pp. 122-25, By to Mann, Mar. 15, 1830. See also W.O.44/18, pp. 164-67, By to Mann, Mar. 15, 1830. By wanted twenty-two blockhouses to defend his canal. He recommended them himself and estimated their cost at £33,000. See also ibid., pp. 153-57, Durnford to Mann, Apr. 24, 1830.
23 W.O.44/18, p. 214, Durnford to Couper, Apr. 24, 1830.

The governor-in-chief at Quebec expressed "surprise and concern" and refused permission to exceed the annual parliamentary grant.[24] But in London the Treasury decided to accept the Ordnance estimate, stating "that it appears to my Lords that there has been some informality in the proceedings of Lt. Colonel By, but that on the whole he has acted correctly. . . ."[25] Despite the fact that four days later the question of By's overspending was raised in the House of Commons,[26] from December 1830 the British government accepted £693,448 as the official sum needed for the canal.[27] Colonel By himself confirmed this figure in a report to the I.G.F. on January 8, 1831.[28] The tone of the correspondence among his superiors, however, now indicated that they had reached the end of their patience. No further upward revisions would be accepted.

The London authorities had been alarmed but tolerant so far. They now turned to open criticism. The Ordnance itself began the attack, sending By his first reprimand since the time of Wellington and Smyth. The Colonel's record was catching up with him. The estimate he had given to the Kempt committee in June 1828 was £576,737, yet in his report of January 8, 1831 he claimed the Kempt figure was his own later estimate of £693,448. Bryce immediately saw the discrepancy and demanded an explanation. The former head of the committee, Sir James Kempt, was the new master general, and he followed Bryce's demand with specific orders:

> In acknowledging the receipt of this letter, Sir A. Bryce will inform Lieutenant-Colonel By, that the Estimate given by him to the Committee in 1828, amounted to £576,757 only, and not, as he states, to the sum of £693,448/11/10. That the latter sum is the amount of his supplementary Estimate, given in by him in 1830, which has never been sanctioned by the Government. That the greater part of money which has been expended, has not yet been voted by Parliament, and that he is on no account to undertake any new work, or to incur any expense in the completion of those now in progress, unless the same should be deemed of pressing importance, and essentially necessary for the due completion and security of the Canal.[29]

Yet the department did not sustain this belated effort to restrain By; other branches of the government took over, and forced the Ordnance to close ranks on the problem of the Rideau Canal and to return to By's support. An attack on the Colonel implicated the department responsible for him, thus the main assault came from the House of Commons and the Treasury.

---

24   Ibid., p. 215, Couper to Durnford, Apr. 27, 1830.
25   W.O.44/20, p. 189, Stewart to Hay, Dec. 2, 1830.
26   The debate is quoted by the *Canadian Courant*, Jan. 21, 1831.
27   P.P. 1830-31 (135) IX, 15, "Canada Canal Communication," p. 133, Treasury minute, by T. Spring-Rice, Feb. 8, 1830, and pp. 114-16, Butler to Stewart, Aug. 17, 1830.
28   P.P. 1831-32 (in 570) V, Report of a select committee of the House of Commons on canal communication in Canada, June 29, 1832, pp. 24-25, By to Bryce, Jan. 8, 1831.
29   Ibid., p. 25, Ordnance minute, by Kempt, 1831.

On March 10, 1831, Lord Althorp, Chancellor of the Exchequer, brought the subject of the Rideau Canal to the attention of the House once again, asking "whether it would not be better to sacrifice that which had already been expended than now to enter on a fresh expenditure," and putting forward the motion that the decision on further spending should be left to a select committee. In the ensuing debate a number of members condemned the whole policy of paying for projects of this kind in the colonies, but Henry Goulburn, Spring-Rice, Henry Labouchere and Sir George Murray, while regretting the large increase in costs, argued that the government was bound to complete the work. In the end the House appointed a committee[30] which shortly afterward drew up four resolutions designed to tighten control over defence spending in the colonies. It recommended the adoption of the following regulations: First, it was deemed that no public work in the colonies could be undertaken without officially approved plans and estimates. Second, no money could be requested from Parliament for a project without prior submission of the most accurate estimates possible to the House of Commons. Third, the House had to be promptly informed of sums spent and sums necessary in case of projects taking more than one year to complete. Fourth, no government department could empower any of its officers to make contracts exceeding the limits of annual parliamentary grants without the specific approval of Treasury.[31] In short, these recommendations were aimed directly at the defects in the imperial financial organization being revealed with increasing clarity in the execution of the Rideau project.

On June 10, 1831, the Treasury circulated the four resolutions to the Colonial Office, Admiralty and Ordnance, asking for comments. The Colonial Office and Admiralty accepted them without qualification,[32] but the Ordnance objected. The master general and board took exception to the fourth resolution on the grounds

> that, unless Executive Officers at *Foreign Stations* are authorized to enter into a contract for the *whole* of any Work proposed to be executed in a given number of years, Contractors will naturally seek to indemnify themselves, by higher prices, from the risk they would run of having their work stopped, or discontinued, at the end of any year. At the same time, the Treasury, in the event of Resolutions being adopted, having, as it seems, the power of sanctioning, by their Minute, a deviation from the 4th Resolution, it is submitted as a question for their Lordships' consideration, whether this power will be effectual to obviate the inconvenience above anticipated.[33]

Perhaps this Ordnance effort to retain the power which gave it maximum financial independence in the colonies was justified in normal circumstances. A department with

---

30 Quoted in the *Brockville Recorder*, Apr. 14, 1831.

31 P.P. 1831-32 (in 570) V, Report of a select committee of the House of Commons on canal communication in Canada, June 29, 1832, p. 36, Treasury minute, June 10, 1831.

32 Ibid., p. 37, Hay to Stewart, June 29, 1831, and Barrow to Stewart, June 17, 1831.

33 Ibid., pp. 37-38, Byham to Stewart, July 15, 1831.

military responsibilities needed the financial flexibility to provide defence works cheaply and quickly, especially in emergencies. Colonel By, however, had demonstrated how the power to make contracts could be abused beyond reasonable limits. As a result the Lords Commissioners overruled the Ordnance objections and ordered that in future all four resolutions were to be regarded as departmental regulations.[34]

The House of Commons and Treasury had taken an important step in restraining Ordnance expenditures. Still, Colonel By remained free to extract more unexpected sums from Parliament. When he proceeded to exploit his freedom further, the Treasury found it necessary to coerce the Ordnance more drastically.

On July 15, 1831, Colonel Durnford forwarded By's latest cost calculations and progress report to London. Durnford stated he had gone over the whole line of the canal with By and had concluded that the work was being carried forward with maximum efficiency and that the additional sums required for its completion were fully warranted. Those additional sums came to £26,626 in excess of the previous estimate, and might still have to be revised upwards.[35] The report reached London in September 1831. On November 4, inspector general Bryce commented that on the whole By's new calculations appeared to be perfectly sound but that some of the additional expenditures nominated as absolutely necessary seemed inadequately explained. Bryce concluded by arguing that ". . . when it is considered that a work of such magnitude and novelty as the Rideau Navigation, has been carried on, not in a settled country, where all the localities and resources might be intimately known, but pushed on by the greatest exertion, in a new country, with a new establishment formed on the spot, it is not surprising that errors should have arisen. . . ."[36] The Ordnance stood by its man on the Rideau. This the Treasury was no longer willing to tolerate.

When the report with Bryce's comments reached the Treasury more than three months later,[37] the response was sharp. The Lords Commissioners called together a second House of Commons investigating committee, and the House ordered all Rideau Canal documents to be placed before it.[38] On February 7 the Treasury rebuked the Ordnance, considering the latest unauthorized rise in the estimate "a matter no less of regret than of surprise." The Lords Commissioners wanted to know why an additional year's work was now needed to finish the canal, and made plain their intention not to sanction any additional expenditures whatsoever without fuller explanations than those they had so far received. They also asked why the Ordnance had been so slow in passing on By's latest figures, and warned that if it were discovered that anyone in Canada had

34   Ibid., p. 38, Treasury minute, July 8, 1831.
35   Ibid., pp. 25-27, Durnford to Fanshawe, July 15, 1831.
36   Ibid., pp. 27-28, Ordnance minute, by Bryce, Nov. 4, 1831.
37   Ibid., p. 19, Ordnance memorandum, by Ellicombe, Feb. 3, 1832. It was only on February 3, 1832 that the Ordnance informed the Treasury of By's latest estimate. In other words, it took three months to send a document from 84 Pall Mall to Whitehall.
38   W.O.44/24, p. 524, Spring-Rice to the Ordnance, Feb. 4, 1832.

departed from the four recently imposed financial regulations it would be necessary to consider serious punishment.[39]

The Ordnance replied three days later. Colonel C. G. Ellicombe referred the Lords Commissioners to By's reports and letters, quoting from them liberally. The original estimates had been inaccurate of necessity because they had been made in a densely wooded, swampy, fever-ridden and sometimes frozen wilderness. It was therefore more surprising that the errors in estimates had proved to be so small rather than that the calculations had been revised upwards. The essence of Ellicombe's memorandum was that the Ordnance could give the Treasury no better explanations than those supplied by Colonel By and that they were considered adequate. Concerning delays in forwarding By's last report, Ellicombe stated that the report had been under close Ordnance scrutiny between September 3, 1831—when it arrived—and November 4; he did not say, however, why it was kept from the Lords Commissioners until February 4 of the following year. The final point of the memorandum was the most curious of all. In Ellicombe's words,

> In reply to the Board's question, as to what communication has been made to Canada regarding the Vote of 1831, I have to observe, that no official communication has been made to this Office of the Vote being granted, and consequently no communication has been made to Canada from hence; indeed as the Vote is taken by another Department, and not by the Ordnance, it is to be presumed that that Department may have made the necessary communication to the Ordnance or proper Authorities in Canada, with a view to check the issuing of any sums beyond the sum voted. . . .[40]

In other words, the Ordnance was asserting that, because the canal was paid for under the Colonial Office budget, financial control over By was really none of its business. The Colonial Office was responsible for the annual expenditure on the Rideau Canal. This assertion marked the extreme point in the confusion over the management of imperial funds overseas. The Treasury refused to accept it as a valid argument.

Ellicombe drafted a second memorandum on February 15, going over the same ground in a more defensive tone and adding the now familiar ultimatum; if the Treasury refused to sanction all of By's financial arrangements, the costs of the canal would rise still higher.[41] On February 20 Richard Byham forwarded both documents to the Treasury,[42] where Ellicombe's arguments were demolished point by point. The Lords Commissioners agreed only that "they cannot avoid submitting an estimate for the Rideau Canal to Parliament" because they were bound by Colonel By's latest financial commitments.

---

39 P.P. 1831-32 (in 570) V, Report of a select committee of the House of Commons on canal communication in Canada, June 29, 1832, pp. 18-19, Treasury minute, Feb. 7, 1832.

40 Ibid., pp. 29-30, Ordnance memorandum, by Ellicombe, Feb. 10, 1832.

41 W.O.44/20, pp. 165-69, Ordnance minute, by Ellicombe, Feb. 15, 1832.

42 P.P. 1831-32 (in 570) V, Report of a select committee of the House of Commons on canal communication in Canada, June 29, 1832, pp. 28-29, Byham to Stewart, Feb. 20, 1832.

In their refutation of Ellicombe's defence, the Lords Commissioners stressed two matters. They condemned the Ordnance for delaying the transmission of financial information necessary for the preparation of accurate requests for funds from Parliament. Such unjustifiable delays could not be tolerated. In the second place, they rejected the Ordnance plea of ignorance about the amounts of the annual parliamentary grants, stating that ". . . there was no necessity for any communication to be made from this Board of the amount voted by Parliament for the purpose. The Votes of Parliament are of themselves sufficient information upon the subject, and it appears to My Lords to be the duty of the respective Departments to take cognizance of the Votes for all Services conducted under their direction and to issue proper instructions upon them."[43] The Lords Commissioners did not consider it their duty to provide Colonel By with financial instructions. It was the duty of the Ordnance to instruct him.

The Treasury was becoming aware of the gap existing between its concept of the proper management of expenditures abroad and the actual working of Britain's finances. When Colonel By submitted his final set of figures, the situation was fully clarified. Belatedly the Lords Commissioners took drastic remedial action.

By drew up his last annual account on February 23, 1832; his costs were up once more. In 1831 he had submitted an estimate for £720,075 to finish the canal, but now he wanted £803,744.[44] To the end of 1831 he had already spent £714,409, which, according to Ordnance calculations, was £22,743 in excess of the parliamentary grants for the period through 1831. In February 1832 the Lords Commissioners had made it clear that they were not prepared to authorize any funds above those granted early in 1832, yet the Colonel was requesting £88,366 in excess of the as yet unauthorized total of £715,409. He seemed quite oblivious to the fact that in the eyes of the Treasury and House of Commons his financial arrangements constituted defiance of higher authority. Instead, he urged his superiors in Canada to complete all the canals from Montreal to Lake Erie on the enlarged Rideau Canal scale. This great seaway could be built in just eighteen months, and would be of such value to the empire that the money needed for it should be no object.[45]

Four days after May 21, 1832, when the Ordnance forwarded By's final estimate to the Treasury,[46] the Lords Commissioners took strong action. Concerning Colonel By, they considered

> It is impossible for My Lords to permit such conduct to be pursued by any public functionary. If My Lords were to allow any person whatever to expend with impunity, and particularly after repeated increases of the original Esti-

---

43   Ibid., pp. 30-31, Treasury minute, May 11, 1832.

44   Ibid., pp. 34-35, Abstract of Rideau Canal estimates and expenditures, by Colonel By, Feb. 23, 1832.

45   W.O.44/20, pp. 33-39, Ordnance minute, 1832, and pp. 97-98, By to the R.O. Quebec, Apr. 10, 1832.

46   P.P. 1831-32 (in 570) V, Report of a select committee of the House of Commons on canal communication in Canada, June 29, 1832, p. 32, Byham to Spring-Rice, May 21, 1832.

mate, upon any work under his superintendence a larger amount than that sanctioned by Parliament and by this Board, there would be an end of all control, and My Lords would feel themselves deeply responsible to Parliament. They desire, therefore, that the Master General and Board will take immediate steps for removing Colonel By from any further superintendence over any part of the Works for making Canal Communications in Canada, and for placing some competent person in charge of those Works, upon whose knowledge and discretion due reliance can be placed; to whom must be furnished a statement of the Estimates and Grants, and who must be strictly charged upon no account whatever to exceed the amount of the Grants.

My Lords further desire that Colonel By may be forthwith ordered to return to this country, that he may be called upon to afford such explanations as My Lords may consider necessary upon this important subject.[47]

The Lords Commissioners not only denounced By, but, by implication, his whole department. They found it necessary to intervene in the internal affairs of the Ordnance, to discipline an officer of another department of state.

The Ordnance did not resist. It signified compliance to the wishes of the Treasury, "lamented" the fact that By had gone so far beyond the bounds of acceptable financial procedures, and promised to adhere to all future directives from the Lords Commissioners. No further Ordnance attempts to defend the Colonel were made.[48] In London all parties were now aware of the extent to which the management of Rideau Canal expenditures had escaped from the grasp of those directly responsible to the House of Commons. A single officer in the Canadian wilderness had been levying taxes on the British public without serious question or restraint from 1826 to 1832. The 1832 Commons committee on Canadian canals now had to discover how this had been possible.

Among the twenty-six members of this committee—a group similar to the 1831 committee on Canadian canal expenditures—were the leading reformers of the day. There was Thomas Spring-Rice, secretary to the Treasury; John Charles Spencer, Viscount Althorp, Chancellor of the Exchequer; Henry Goulburn, a man of wide experience in both the Treasury and Colonial Office; Henry Labouchere, a strong supporter of economic reforms; Edward George Geoffrey Smith Stanley, another strong liberal reformer; and, above all, Viscount Howick (later Earl Grey), one of the most active advocates of economy in the colonial administration. Sir George Murray, former governor of Canada and currently master general, and Sir Henry Hardinge, probably second only to Wellington among Britain's soldiers, represented the military interests. Aside from the soldiers, these men were all devoted to efficiency and economy within the imperial government. The Rideau Canal problem was only one among many, and they could devote relatively little time to it. But it probably helped to reinforce their attitudes towards colonial military expenditures. Howick and Labouchere later both

---

47   Ibid., p. 33, Treasury minute, May 25, 1832.
48   Ibid., pp. 39-40, Ordnance memorandum, May 26, 1832.

became colonial secretaries, the former between 1846 and 1852 and the latter between 1855 and 1858. Both helped to reduce Britain's military commitments in the colonies. Grey initiated the policy of withdrawing the legions and among the first military branches he removed from the colonies was the Ordnance.[49]

The committee's investigation revealed many administrative defects without finding answers to most of its own questions. In fact, the failure to find these answers itself revealed the lack of control over colonial finances and officers abroad. The key witness, Colonel By, did not appear, nor did many of the most important Rideau Canal documents.

Almost at the same time the Treasury demanded his recall, By—with his family and officers—was celebrating the opening of his canal to the applause of the inhabitants of Kingston.[50] From there he immediately set out for England, authorized to come home by the Ordnance on his own request.[51] He was neither dismissed nor recalled, but on his way home at the time the House of Commons committee concluded its proceedings. By did not become fully aware of the Treasury censure until more than a year later.[52]

As for the documents available to the committee, these were voluminous but incomplete. Letters and memoranda containing By's visionary schemes costing millions, the criticisms of Wellington, Smyth, Anglesey, Huskisson and Canadian governors, and many other items were missing.[53] Without these a full evaluation of the Colonel's activities was impossible.

On June 15 the committee called in three witnesses, Colonel Durnford, Colonel Ellicombe of the inspector general's office, and William Sargent, superintendent of the Commissary Department. Durnford, technically By's immediate superior in Canada, was questioned first. The committee members were primarily interested in the degree to which Durnford was responsible for the management of the Rideau project. He answered all their variously phrased questions the same way; By had been "totally independent of the command I held in the country." Durnford, whose office was at Quebec, had visited the canal site only occasionally and unofficially. Colonel By had

---

49   House of Commons Journal, 1831-32, Vol. 87, p. 368. On Viscount Howick's economic reform activities, see Peter Burroughs, "The Search for Economy: Imperial Administration of Nova Scotia in the 1830's," *C.H.R.,* XLIX (1968), 24-43.

50   Kingston *Chronicle,* May 26, 1832. The editor described the arrival of By and his entourage from Bytown on the first steam vessel to pass through the full length of the Rideau Canal.

51   W.O.44/16, p. 5, Bryce to Couper, May 30, 1832.

52   Ibid., pp. 13-19, By to Pilkington, July 22, 1833.

53   The 1832 committee had two sets of printed papers for examination. The first, P.P. 1830-31 (135) IX, 15, "Canada Canal Communication," consisted of 133 folio pages containing copies of 131 documents and an appendix of six plans. The second, P.P. 1831-32 (in 570) V, Report of a select committee of the House of Commons on canal communication in Canada, June 29, 1832, was a set of 38 pages of documents and one map, appended to the committee's report. Possible explanations for the omission of many important documents are inefficiency, deliberate editing on the part of an alarmed Ordnance, and the private or confidential nature of some of the papers.

operated outside the normal chain of command. Durnford added that from what he had seen he considered By's work and judgement sound and proper. When asked if he had considered the possibility of building the whole Ottawa-Rideau waterway on the small scale, Durnford answered he had "always" believed it was the government's intention to build the canals on By's larger scale. Such a belief could only have come to him from By and indicates the possibility that By had "always" intended to build a large scale canal. The committee, however, did not follow up this line of inquiry but turned to Ellicombe, who contributed little to the evidence.[54]

Sargent's testimony, on the other hand, revealed in detail the shortcomings in the imperial system of budgeting, accounting and audit. According to him, the branches of the Commissariat in the colonies, entrusted with the payment of all money for military needs, disbursed funds without knowing how much Parliament had allocated to military purposes in any given year. The Treasury, of which the Commissariat was a branch, did not normally inform its own overseas representatives of the amount of annual parliamentary votes. There was no effective control over Commissariat spending except in special instances where the Treasury gave the Commissariat direct orders concerning annual limitations on expenditure. In the Rideau Canal case the Treasury had given no orders and therefore the Commissariat in Canada had followed normal procedures and given By whatever he requested. Indeed, the Commissariat had no power to refuse requests for funds from Ordnance officers overseas whether these funds exceeded parliamentary budget allocations or not. It had no part whatsoever in drawing up or checking the accounts of such officers.[55]

Consequently, in its report the committee stated that while there was still no way of knowing what the final canal cost would be "the Public must be ultimately liable for the payment of the whole" because of By's entangling contracts and independent powers to commit his government to whatever expenditures he felt were necessary. Even Parliament was powerless to stop the unauthorized expansion of spending on a public work once it was in the construction stage. As for accounting and audit restraints, it seemed there was no provision for them. Colonel By's accounts were repeatedly examined by the Ordnance, but always too late to alter them and never with sufficient information from sources other than By himself to enable a critical evaluation. Machinery for external audit was nonexistent. The Commons committees of 1831 and 1832 had to attempt the audit of By's calculations themselves—without access to basic documentation—at a time when they could no longer prevent a large misallocation of public funds or even establish clearly the reasons for the misallocation.

The 1832 committee could only recommend that steps should be taken to prevent a recurrence of such a financial disaster. The members repeated their various "regrets"

---

54   P.P. 1831-32 (in 570) V, Report of a select committee of the House of Commons on canal communication in Canada, June 29, 1832, pp. 7-13, Examination of Durnford and Ellicombe, June 15, 1832.
55   Ibid., pp. 7-13, Examination of Sargent, June 15, 1832.

and expressed surprise that the recommendations of the 1831 Commons committee, converted into regulations by the Treasury, had exercised so little corrective influence. However, "By a Treasury Minute of May 11, 1832, some restraint is imposed on the application of Money by the Commissary General of Canada. Your Committee have observed this with pleasure, being of opinion, that in Works of this description there is no security against extravagance; and the amount of expense can never be accurately known, unless the annual expenditure be confined within the limits of the annual Votes."[56]

The House and the Treasury also moved to cut back other colonial defence projects which might get out of control. They began to choke off funds for Ordnance works abroad early in 1830. On January 17, 1831, the Ordnance summarized the financial status of its major fortifications undertakings. By that date only the defence works on Corfu in the Ionian Islands were completed; they had been paid for by the local government. In British North America Treasury and Commons had provided only £50,427 out of £214,863 for Halifax and even the works at Quebec, begun by Richmond in 1819, were not yet complete because only £222,399 out of £236,540 had been allocated to them. Progress on the fortifications at Kingston, begun at the same time as the Rideau Canal, was especially slow, with just £40,000 out of £273,000 provided so far.[57] In January 1832, when the Rideau Canal had become the most expensive Ordnance project in the empire, Sir Alexander Bryce was forced to eliminate a number of essential outworks planned for the Kingston fortifications through lack of funds.[58]

The House committee of 1832 ordered further economies. To eliminate canal maintenance costs it urged that the completed military waterway should be turned over to the management of the Canadians. All outstanding Ordnance plans to complete the Ottawa-Rideau navigation on By's enlarged scale were also vetoed. Because part of the estimates for these additional works were framed by Colonel By, the committee did "not think that it would be prudent to rely on their accuracy." Plans to widen the three lower locks of the Grenville Canal, completed on the small scale before By altered the width to thirty-three feet, were stopped. Nothing more was to be spent on Canadian military canals.[59] Therefore, only military or naval vessels of less than twenty foot width could pass all the way from Montreal to Kingston by this route. A Rideau Canal on the original scale would have served just as well, and from London's viewpoint By's massive expenditures had been a waste.

Having thus punished the Ordnance and crippled Wellington's Canadian defence system, the guardians of the public purse turned to By. The Colonel returned to England expecting some public honour for his great achievement. Early in 1833 he

---

56   Ibid., pp. 2-6, The report of the committee, June 29, 1832.
57   W.O.44/19, p. 219, Ordnance minute, Jan. 17, 1831.
58   W.O.44/32, p. 21, Ordnance memorandum, by Bryce, Jan. 19, 1832.
59   P.P. 1831-32 (in 570) V, Report of a select committee of the House of Commons on canal communication in Canada, June 29, 1832, pp. 2-6, The report of the committee, June 29, 1832.

applied to the master general for an official reward, and the head of the Ordnance considered it would be "very proper" to honour him by presenting him at His Majesty's Levee.[60] Pressure from high places swiftly led to the abandonment of this idea. But By continued to expect a reward even when the Treasury began to assail him, through the Ordnance, with a series of demands for explanations of his financial proceedings.[61] He consistently answered that all his transactions had been justified, but in marshalling supporting evidence he once more demonstrated a talent for dissembling.

He submitted an abstract of Rideau expenditures on January 12, 1833, in which he put the final canal cost at £788,286.[62] Yet a year before he had presented his successor on the Rideau, Captain Donald Bolton, R.E., with a document in which the final cost was £803,744.[63] Then, on January 18 By claimed he was waiting for Bolton to tell him the final figures.[64] On February 14, 1833, he concluded that the true figure was £777,146.[65] According to Ordnance calculations made in 1834, however, he actually spent £822,804.[66]

In an attempt to explain why his estimates had constantly risen he produced self-incriminating arguments. His first complete estimate of November 1827, amounting to £474,899, had been deliberately calculated as the absolute minimum because

> ... there was such a vast difference of opinion between Sir J. C. Smyth and myself relative to the expense of constructing the Rideau Canal, I felt it my duty to form the Estimate so as to shew [sic] the least possible expense that the Rideau Service would require; and the Estimate given to the Committee in June 1828 amounting to £579,000—was formed on the same principle; I therefore most respectfully trust that when these circumstances are taken into consideration, it will be a matter of no surprise that the total amount has nearly doubled the original Estimate for small Locks.[67]

In short, By admitted that out of a sense of duty he had drawn up estimates intended to mislead his superiors.

About June 22, 1833, By finally received an Ordnance copy of the Treasury minute of May 25, 1832, ordering his dismissal and recall. In shock and indignation he protested total innocence of all implied Treasury charges. He claimed that he had never spent an unauthorized penny; if some of the funds he had used in strict compliance with orders to work at maximum speed had not been properly allocated to his project, that was the fault of his superiors. Each of his annual accounts had been approved by the governors and lieutenant governors of Canada as well as by his local commander, Colonel

---

60 W.O.46/32, p. 186, Couper to Pilkington, Feb. 12, 1833.
61 W.O.44/15, p. 268, Stewart to the Ordnance, May 11, 1833.
62 Ibid., p. 255, Abstract of Rideau Canal estimates and expenditures, by Colonel By, Jan. 12, 1833.
63 Ibid., p. 130, Bolton to the R.O. Quebec, Nov. 6, 1832.
64 Ibid., pp. 258-60, By to Fanshawe, Jan. 18, 1833.
65 Ibid., p. 264, By to Fanshawe, Feb. 14, 1833, and W.O.44/21, pp. 358-59, By to Ellicombe, Dec. 20, 1833.
66 W.O.44/15, p. 297, Ordnance statement of Rideau Canal costs, by S. Thomas, Jan. 29, 1834.
67 Ibid., pp. 272-79, By to Ellicombe, May 29, 1833.

Durnford. Of course these assertions do not stand up because Canadian governors had ordered him not to exceed annual parliamentary grants, and Durnford, by his own testimony, had never exercised authority over Rideau Canal expenditures.

By concluded his protest by stating he had built his great work more quickly and cheaply than any comparable project in the past, and that

> I feel myself utterly guiltless of any wilful dereliction of duty as an Officer and a Soldier, and it therefore cannot surprise anyone that I should feel the deepest mortification at the condemnation of my conduct as recorded in the Treasury Minute already referred to. Up to the period when this paper was communicated to me, I conceived that I was so discharging my duty as to entitle me to the praise and approbation of my Government rather than its censure. . . . [68]

To expunge the stigma placed on him by the Treasury he again requested the award of some public mark of distinction, presumably a knighthood.

The Lords Commissioners were unmoved. Instead, they now accused him of conduct unbecoming an officer because he had constructed two private residences on government land at Bytown on the pretext that the land belonged to him, and had indulged in real estate transactions involving other portions of public property.[69] The Colonel, deeply wounded, protested that all his land dealings had been in the public interest.[70] In fact he had spent over £800 of his own money to house his family and Lieutenant Pooley, and the houses had now been taken over by the government. Rather, he wanted compensation.[71] The Lords Commissioners relented and withdrew their accusations concerning illicit real estate transactions, but they refused to compensate him for his houses.[72]

Until the end of 1834 the Colonel continued to pen letters defending his conduct, giving advice about the management of the Rideau Canal and urging more Canadian canal projects.[73] On October 18, 1834, however, he suffered a serious attack of apoplexy and paralysis. According to his physician, Dr. N. J. Thomas, inspector general of hospitals, the attack was brought on by anxiety, and other attacks would follow unless his peace of mind was restored.[74] On Christmas Day, 1834, By's wife, Esther, appealed to the Ordnance to save her husband's life by giving him some public honour. She considered the Treasury minute of May 25, 1832, directly responsible for his illness.[75]

---

68 W.O.44/16, pp. 13-19, By to Pilkington, July 22, 1833.

69 W.O.44/15, pp. 360-63, Treasury minute, Sept. 10, 1833.

70 Ibid., pp. 364-71, By to Pilkington, Nov. 22, 1833.

71 Ibid., p. 355, By to Pilkington, Sept. 9, 1833.

72 Ibid., p. 372, Stewart to the Ordnance, Dec. 31, 1833, and p. 374, Stewart to the Ordnance, Jan. 27, 1834. There were also later appeals made in the Colonel's behalf for compensation. The Treasury again denied one of these in May 1835 (see ibid., p. 380, Stewart to the Ordnance, May 5, 1835). In 1849 the executors of By's will also petitioned the Ordnance for compensation in behalf of his heirs. The Ordnance could not help (see W.O.44/47, Ordnance minute, by Thomas Clarke, Dec. 21, 1849).

73 W.O.44/42, p. 211, By to Fanshawe, Aug. 22, 1834.

74 W.O.44/16, p. 39, Dr. N. J. Thomas to Mrs. By, Oct. 25, 1834.

75 Ibid., pp. 29-31, Mrs. By to the Ordnance, Dec. 25, 1834.

The inspector general of fortifications was sympathetic but could not help.[76] Colonel By lingered on painfully for another year and then died on February 1, 1836, at the age of 57.[77] If Dr. Thomas and Mrs. By are to be believed about the effect of the Treasury minute, John By was punished more severely than anyone intended.

The treatment he received seems especially harsh because no one in London was prepared to weigh his technical achievements against his administrative sins. He had overcome formidable and even unique constructions problems on the Rideau waterway. At the commencement of the project he was faced with a 123-mile long chain of rivers, lakes and creeks flowing in various directions at different elevations through a heavily forested and almost entirely uninhabited wilderness. Initial survey parties, travelling by canoe, were forced to start clearing some of the bush and no construction was possible until large numbers of trees and undergrowth were removed. Aside from a few oxen the whole waterway had to be built by manual labour with manpower imported from distant settled areas. Timber, stone and earth were locally available, but manufactured items such as tools and steel and iron had to be brought from as far away as England. In these respects the Ottawa canals and the Lachine and Welland canals were easier to build; in general they followed settlement while the Rideau Canal preceded it. Furthermore, the Colonel's military workforce—reliable and expert—consisted only of ten Royal Engineers and two companies of Sappers. He therefore had to rely for the bulk of the work on civilian contractors, many of whom had little or no experience and ultimately failed to meet their obligations; the result was wasted work, delay, additional expense, and much irritating litigation. The bulk of the 2,000 or more civilian craftsmen and labourers were recently-arrived Irishmen, unaccustomed to the extremes of the Canadian environment. On the numerous swampy sections of the canal route a form of malaria called "swamp fever"—to which Indians and French-Canadian voyageurs seemed almost immune—took a heavy toll of life among these men in the summer. In 1828 the disease almost halted construction and By himself almost died from it. Finally, the technique used to form the waterway involved extensive flooding by means of dams in order to raise water levels to the required depth between the series of locks, and the local owners of the as yet unsettled lands made serious difficulties for the Colonel, asking unreasonable prices and annoying him with lawsuits.[78]

During construction there were recurring engineering difficulties. In the autumn of 1826 the Colonel began work by throwing bridges across the Ottawa near its juncture with the Rideau; the longest of these collapsed and had to be rebuilt. Early in 1827 he set up a headquarters, barracks, a hospital, and workshops at what became Bytown, and worked on the first eight locks leading up from the Ottawa River while other parties began construction at the Kingston end. These first locks, built of unbonded limestone

---

76 Ibid. The inspector general's reactions to this letter are inscribed in a marginal comment on the letter itself, dated Dec. 29, 1834.

77 Legget, *Rideau Waterway*, p. 61.

78 Ibid., pp. 34-55.

blocks, leaked badly; a technique had to be worked out whereby locally manufactured liquefied cement was forced into the masonry under pressure. All the remaining locks required this treatment, increasing costs and construction time. The next and most dramatic problem occurred five miles up the route at Hog's Back, where a large dam was needed. The contractor in charge used inappropriate methods and the 1828 spring floods washed the 35-foot high structure away; the contractor gave up and departed. This failure embarrassed the Colonel and he took pains to justify the unfortunate circumstances to his superiors. But they were preoccupied with his costs and accepted this engineering failure without serious question, one officer noting that after all the Deep Cut on the Welland Canal had also collapsed.[79] The task was given to the Sappers but they too failed. Finally Captain Victor, R.E., redesigned the plans and the dam was successfully completed early in 1829. Further towards Kingston, at Nicholson's Rapids, a serious error had to be corrected involving the relocation of one of the two locks, an expensive task. Similarly the original plan at Merrick's Mills (Merrickville today) had to be altered and a major mistake was made at Mud Lake, where a simple dam could have avoided the cutting of a costly mile-long canal. At Jones Falls it was necessary to construct the biggest dam on the waterway, 65 feet high and 130 yards long at the top. The project was approached with some anxiety but contractor John Redpath executed it without trouble, making it the most impressive single work on the canal. As well, there were minor problems such as the lock gate opening and closing mechanisms which had to be completely redesigned. Direct supervision of the various parts of the canal and some responsibility for these problems was in the hands of the contractors and By's engineer officers. The ultimate responsibility and the final achievement, however, were his.[80]

The finished product was an outstanding engineering accomplishment. A total of 33 solid masonry locks raised the levels of the various bodies of water 277 feet above the Ottawa River to the watershed at the Upper Rideau Lake and 14 lowered the levels by 162 feet to Lake Ontario at Kingston; they allowed passage of vessels 110 feet long, 30 feet wide, drawing $5\frac{1}{2}$ feet of water. In addition to numerous dams and sluice gates to control the depths of natural watercourses there were 18 miles of artificially excavated canals, parts of which were carved out of solid rock. These were 60 feet wide at the bottom and 80 feet wide at the top.[81] In 1832 the Rideau Canal was by far the largest in Canada and one of the largest in North America. The Erie Canal was much longer but a narrow and shallow work suitable for barges rather than steam vessels; and the Rideau Canal was better constructed, of more durable materials. Captain Richard Bonnycastle, R.E., himself one of the canal builders, expressed the sentiments of many contemporary

---

79 P.P. 1830-31 (135) IX, 15, "Canada Canal Communication," p. 76, By to Mann, Apr. 4, 1829; p. 77, Kempt to Murray, Apr. 30, 1829; By to Couper, Apr. 23, 1829.

80 E. C. Frome, "Account of the Causes which led to the Construction of the Rideau Canal . . . ," *Papers and Subjects connected with the Duties of the Corps of Royal Engineers,* London, 1937. See also Appendix B.

81 Legget, *Rideau Waterway,* p. 17.

observers when he wrote that "if ever any man deserved to be immortalized in this utilitarian age, it was Colonel John By"; the canal By built was "perhaps one of the finest works of the kind in the world. . . ."[82]

Yet this great work was completed by an officer who could not be restrained by London, whose activities showed the damage a minor imperial proconsul could do by exploiting a defective system of colonial administration. The total harm By inflicted on his government was, of course, relatively small and quickly forgotten. Yet for the Ordnance in Canada his canal became the focus of all the department's difficulties.

---

82   Richard Henry Bonnycastle, *The Canadas in 1841*, II, 67-68.

Chapter VIII

Conflict Over Ordnance Property

In terms of scale, durability, military importance and civilian utility, the Rideau Canal was the greatest Ordnance achievement in Canada. It was also the most expensive, and as a result, the department had to complete all the other works of Wellington's defence scheme with little prospect of obtaining the necessary funds from Parliament. Consequently, Ordnance officers began to look to their Canadian possessions as an alternative source of revenue. From the transfer of all military properties to the department in 1826 they acquired large land holdings and the Rideau and Ottawa canals gave them control over a potentially lucrative communications network as well as more land. With skilful exploitation these assets promised to provide the means to secure Canada from invasion without reliance on British money.

Canadians, however, disliked the presence of an alien landlord holding thousands of acres in many of their fastest growing communities. They resented the tolls and other burdens imposed on them since 1832 for the use of the Ordnance canals. Thus land disputes began while the Rideau Canal was still incomplete, and intensified thereafter. Every Ordnance attempt to raise revenue was met with more opposition, with the Colonial Office and its representatives supporting the Canadians. The colonists even refused to recognize the department's legal right to hold property. After a prolonged struggle the Ordnance did gain full title to its possessions in 1843, but the effort to raise funds, especially from the large Rideau assets, failed. By 1845 this constant conflict had exhausted the department's patience and energy, curtailing all of its activities.

While the official transfer of military property to the Ordnance took place in 1826, in practice the department encountered difficulties even before Canadians and their governors attacked its claims. Information as to the extent, location and boundaries of military lands in both Canadian provinces was incomplete. In 1825 the master general and board ordered the respective officers in Canada to clarify the confusion and to assemble all documents necessary to establish clear Ordnance title.[1] The officers, with

---

1   W.O.44/40, p. 92, Ordnance circular to all R.O. at foreign stations, by W. Griffin, June 9, 1825.

insufficient personnel, had only a limited success. A full record of Ordnance lands was not compiled until the late 1840s. Thus disputes over titles and boundaries were inevitable from the start.

The first notable conflict began in May 1827. William Forsyth, a hotel keeper, fenced off a section of the Niagara River shore in front of his property in order to monopolize the best view of Niagara Falls, already an important tourist attraction. The whole river bank facing the United States, including the fenced-off land, belonged to the Ordnance. Two neighbours competing with Forsyth in the hotel business complained about the fences to Sir Peregrine Maitland, the lieutenant governor; Maitland referred the matter to Captain George Phillpotts, commanding Royal Engineer in Upper Canada. The Captain asked Forsyth to remove all fences and buildings from Ordnance property. When this produced no results Phillpotts assembled a small detachment of troops, and, in the presence of a civilian surveyor and the local sheriff to ensure all proprieties would be observed, ordered the soldiers to tear down everything Forsyth had erected on military land. The hotel keeper protested but there was no violence. As soon as the Captain departed, however, the fences went up again, and the troops returned to tear them down a second time. Forsyth then took Phillpotts to court for alleged illegal entry but lost the case. Undaunted, he appealed to the legislative assembly, charging the Royal Engineer with forceable destruction of his property.

The assembly took up the hotel keeper's cause for political reasons, accusing Phillpotts and Maitland of military despotism and coercion of a civilian. When Maitland refused to allow two of his officers to appear as witnesses in front of an assembly investigating committee, he was further charged with denial of civil rights and free speech. The battle did not last long in Upper Canada as Maitland and the assembly engaged in political disputes of other kinds, but the so-called Phillpotts case itself went to the imperial parliament where two Commons select committees investigated the Canadian accusations against the lieutenant governor.[2] This episode was indicative of the potential conflict which the question of Ordnance lands could arouse.

In the same year another dispute occurred in Lower Canada involving Mrs. Charles William Grant, Baroness of Longueil. The Baroness formally protested to the governor-in-chief, Dalhousie, that Ordnance officers were attempting to rob her of her lands. Dalhousie was sympathetic and appealed to Viscount Goderich, the colonial secretary, expressing strong feelings against Ordnance interference in provincial affairs. He pointed out he had no power to control Ordnance officers; "Their correspondence is unknown to me, and their acts equally withheld from the knowledge of the Governor or General Officer Commanding the Forces." The land claims of these officers were "fraught with incalculable mischief, alarming every man of property at the idea of being

2   P.P. 1833 (543) XXVI, 229, p. 1-27, Report of two Commons select committees on the Phillpotts case, July 18, 1833, pp. 2-6, Chief Justice J. B. Robinson to Lieutenant Colonel Rowan, Dec. 31, 1832, and p. 6, Statement by Phillpotts, Jan. 7, 1833. See also Gerald M. Craig, *Upper Canada; The Formative Years, 1784-1841* (Toronto, 1928), p. 192.

placed at the mercy of a Board of inferior Officers," a board which in effect was undermining Dalhousie's authority.[3]

The master general, Anglesey, objected to Dalhousie's criticisms and defended the rights of his department. He wrote that ". . . it is to be observed that the Ordnance are The Guardians of their own Property. That They take measures to protect it, without reference to any one or to any Department, and that if Their inferior executive Officers are to be discouraged from performing the Duties imposed upon Them by The Heads of Their Departments, the Public Interest must materially suffer."[4] In May 1829 Sir John Colborne, lieutenant governor of Upper Canada, challenged the whole policy of transferring military properties from Colonial Office to Ordnance control. He asked the department to turn over to his commissioner of Crown Lands all holdings not immediately required for military use so that they could be leased or rented to provide revenue for the public service.[5] The Ordnance agreed that the leasing idea was sound but wanted to know how the revenues would be allocated.[6] Colborne replied that most of the lands in question had been reserved for military reasons by various authorities so long ago that both the purpose and boundaries of the reserves were now obscure. Since the Ordnance never owned these lands, revenues from them should be carried to the account of the King's casual and territorial fund, and used by the lieutenant governor for primarily civilian purposes.[7] The Ordnance countered by arguing that all military lands were needed in one way or another for defence. The master general, Beresford, insisted that no Ordnance possessions could be given up under any circumstances. The department was currently building very extensive defence works and it was impossible to tell what lands would be needed for them all. The Canadians only wanted to make money from these properties, leaving the British government to bear the whole cost of provincial security.[8]

A leading advocate of taking maximum advantage of the military reserves for Ordnance purposes was Colonel Gustavus Nicolls of the engineers. He became concerned when the Upper Canadian legislature began to grant Ordnance land to companies or individuals intending to make public improvements such as railways, docks and canals. The Ordnance had to pay for every foot of land needed for expensive projects like the Rideau Canal, yet the colonists were taking the department's property without even asking permission. This was unjust and detrimental to the interests of defence. Nicolls therefore proposed a plan of exchange, lease and sale, so that the Ordnance could get the benefits from the reserves before the Canadians seized them.[9] His superiors agreed and

---

3   C.O.42/212, p. 272, Dalhousie to Goderich, July 27, 1827.
4   C.O.42/231, Minute by the master general, Oct. 8, 1827.
5   C.O.42/388, p. 319, Colborne to Sir George Murray, May 21, 1829.
6   C.O.42/390, p. 52, Byham to Hay, July 29, 1829.
7   C.O.42/389, pp. 245-55, Colborne to Murray, Dec. 3, 1829.
8   W.O.44/1488, pp. 328-34, Ordnance minute, by Alexander Bryce, Jan. 13, 1830.
9   W.O.44/29, pp. 474-76, Nicolls to Pilkington, Apr. 20, 1833, W.O.44/42, pp. 395-98, Nicolls to Pilkington, and W.O.44/28, Nicolls to the inspector general of fortifications, Sept. 18, 1834.

arranged with the Colonial Office to sell part of the Ordnance property at York in order to raise money for barracks. The Upper Canadian lieutenant governors consistently opposed this scheme.[10] In 1835 it was finally agreed that the Ordnance would get two-thirds of the proceeds, the rest going to the provincial government.[11] Even then the dispute over the York reserves was not completely resolved.[12]

In the meantime the Canadians already had taken action which posed the most serious threat to the department's possessions. When Ordnance officers attempted to complete the steps necessary to acquire full legal title to all reserves in 1827 and 1828 they discovered an apparently insurmountable obstacle. According to Canadian legal opinion the Ordnance Vesting Act of 1821 did not apply to colonies with representative institutions. The attorney general's department of Upper Canada ruled that, since "the Master General and Board were not a Corporate Body . . . [they] therefore could not legally hold real property."[13] But a number of years elapsed before the Ordnance became fully aware of the importance of this ruling. During the interval the Rideau Canal was completed—and these kinds of disputes intensified.

Prior to 1832 the Colonial Office paid for canal construction; the Ordnance only executed the project. As it neared completion, the London authorities began to worry about maintenance costs, and as early as January 1828 Wilmot Horton proposed to transfer the waterway to the Canadians in order to avoid these expenditures.[14] In December 1830, however, the Treasury decided that the Rideau and Ottawa canals, with all their ancillary properties, should go to the Ordnance instead.[15] Consequently the department ordered its officers to take the necessary legal steps to gain possession.[16] It also asked the Treasury to find out if the colonists would pay the costs of upkeep until such time as revenues from tolls could absorb running costs.[17] The Lords Commissioners passed this request to the Colonial Office,[18] and the latter, while anticipating difficulties, instructed its governors to approach the provincial legislatures.[19] In March 1832 Colborne reported that Upper Canada was not interested,[20] and in May Lord Aylmer, the administrator of Lower Canada, reported the same attitude among the legislators and councillors of his province.[21]

10 C.O.42/418, p. 88, Colborne to Hay, Jan. 16, 1834.
11 C.O.42/428, p. 60, Byham to Hay, Feb. 18, 1835.
12 W.O.44/42, pp. 407-09, R. S. Jameson to John Joseph, July 7, 1836 and W.O.44/29, pp. 363-75, Ordnance minute, by W. Wade, Mar. 18, 1835.
13 W.O.44/28, pp. 113-14, J. R. Wright to Nicolls, July 8, 1834.
14 C.O.42/219, pp. 180-81, Horton to Stanley, Jan. 1, 1828.
15 W.O.44/20, p. 189, Stewart to Hay, Dec. 2, 1830.
16 C.O.42/234, p. 196, Byham to the R.O., Quebec, June 29, 1831.
17 Ibid., p. 196, Byham to Stewart, June 29, 1831.
18 Ibid., p. 192, Spring-Rice to Goderich, Sept. 12, 1831.
19 P.P. 1831-32 (in 570) V, Report of a select committee of the House of Commons on canal communication in Canada, June 29, 1832, p. 17, Hay to Spring-Rice, Sept. 24, 1831.
20 C.O.42/411, p. 109, Colborne to Goderich, Mar. 23, 1832.
21 C.O. 42/236, pp. 351-53, Aylmer to Goderich, May 8, 1832.

While this bad news was on its way to London the 1832 House committee on Canadian canals reverted to Wilmot Horton's idea in its report.[22] Independent of Canadian protests, the Treasury rejected the recommendation. The Lords Commissioners decided the Ottawa-Rideau waterway was too valuable to give away. Influenced by a set of Colonel By's calculations, they concluded the canals would be a great asset instead of a liability.

In November 1830 By had predicted that the annual revenues derived from tolls and duties levied on civilian traffic through the Rideau Canal would amount to £41,763. There would also be "enormous savings" in the transportation costs of military and naval stores. All this money would not only cover maintenance and repair expenses but would shortly restore the total outlay for canal construction to the government.[23] The colonial secretary apparently heard of this calculation either from the Ordnance or By himself, and passed it to the Treasury with assurances that it was reliable. The Lords Commissioners were persuaded;[24] they re-issued their 1831 orders transferring the military canals to the Ordnance, and from 1832 the department took possession of the whole property. It proved to be a troublesome asset.

* * * * *

For a start, the true extent of By's confused land purchases was not known until 1840. The Ordnance had to send out two special commissioners, Richard Eaton and J. S. Elliott, to clarify the situation. Eaton stayed only briefly, but Elliott became the permanent Canadian Ordnance commissioner with extraordinary powers, including the power of attorney to act directly in behalf of the master general and board. Before Eaton returned to England,[25] he and Elliott completed a report on November 9, 1840,[26] containing the first reliable picture of the Rideau land situation. Lord Dalhousie had purchased 415 acres of what became Bytown from Hugh Frazer for £750 in 1826 and had turned this land over to By's control.[27] The Colonel himself had purchased 12,200 acres along the canal route. Of these only 101 acres were actually taken up by the canal and its banks, 1754 acres were under water, 9,533 acres were unimproved, and 832 acres were cleared and suitable for settlement.[28] How and why had all this extra land been acquired?

Despite government efforts to get free right of way for the Rideau Canal, in February 1827 Upper Canada had implemented its Rideau Canal Act, by which Colonel

22 P.P. 1831-32 (in 570) V, Report of a select committee of the House of Commons on canal communication in Canada, June 29, 1832, pp. 2-6, The committee report, June 29, 1832.
23 W.O.44/20, p. 49, By to Durnford, Nov. 26, 1830.
24 Ibid., pp. 130-33, Treasury minute, Aug. 17, 1832.
25 W.O.44/136, pp. 161-84, Minute by Thomas, Dec. 8, 1841.
26 R.G.7, G I, Vol. 47, pt. 2, pp. 455-57, Byham to Stephen, May 6, 1840.
27 National Library of Scotland, Edinburgh, Papers of Sir George Murray, Adv. MSS, 46.1.1.-47.7.4., Vol. 182, p. 136, Byham to Elliott, Mar. 10, 1841, and W.O.55/284, pp. 1-52, Eaton and Elliott report, Nov. 9, 1840.
28 W.O.55/284, pp. 1-52, Eaton and Elliott report, Nov. 9, 1840.

By was empowered to acquire all necessary land through purchase.[29] General Smyth had been disappointed, fearing the Canadians would inflate prices,[30] but By had been satisfied because his freedom of action was increased.[31] This statute could also be construed as a retroactive justification for purchases the Colonel had already made prior to its passage into law. His superiors had expected trouble over these improperly authorized acquisitions.[32] In fact By himself was the first to suffer.

In April 1830 he called for help from the Ordnance because of legal difficulties with the local inhabitants. By explained that he had taken up the lands he thought necessary in 1827 and 1828, had cleared them at government expense and greatly increased their value. The former owners now rejected the prices they had agreed on in 1827, were forceably taking back their lands and reselling them to new settlers for inflated sums. When By tried to resist their illegal actions they threatened him with lawsuits which he could not avoid.[33] According to the chief justice of Upper Canada, John Beverley Robinson, the Rideau Canal Act could only protect the Colonel from the consequences of such suits, not from the suits themselves.[34] As a result, By was subjected to constant, irritating litigation.

His legal problems were complicated because he had taken up more land than he needed. He explained that the canal required only 150 feet of clearance on each side, but he had been forced to clear from 300 to 400 feet in order "to obtain a free circulation of air, which was done at considerable expense with the hope of checking sickness. . . ."[35] This sickness was "swamp fever," probably a form of malaria, prevalent in many parts of Upper Canada during summer. A known cure for it was sulphate of quinine, but the drug was too expensive for By's canal workers. Fresh air was also thought to be a preventative of fevers, considered to be caused by "noxious effluvia" or foul gases emanating from swampy areas.[36] Thus By had cleared the extra land as a health measure. He now proposed to lease this property back to its former owners on condition that they dropped all damage claims associated with canal construction.[37] Neglecting to remind his superiors that in 1827 he had wanted to build a fifty-foot wide canal requiring a proportionately wider clearance, he concluded by asserting that the leasing plan he sought permission for was already implemented.[38]

Colonel Ellicombe of the inspector general's office was not pleased. He stated that no one had given By authority to purchase land before the passage of the Rideau Canal

---

29 *Statutes of Upper Canada,* 8 Geo. IV, Cap 1, A.D. 1827.
30 W.O.44/42, pp. 300-05, Memorandum by Smyth, Dec. 3, 1827.
31 C.O.42/384, pp. 75-76, Maitland to Huskisson, July 7, 1828.
32 W.O.44/20, pp. 297-302, Kempt to Murray, Apr. 2, 1829.
33 W.O.44/15, p. 56, By to Mudge, Apr. 14, 1830.
34 Ibid., p. 42, Robinson to Hagerman, 1830.
35 W.O.44/20, pp. 234-35, By to Durnford, May 4, 1830.
36 Leggett, *Rideau Waterway,* p. 52.
37 W.O.44/15, p. 50, By to Mann, Apr. 3, 1830, and p. 53, By to the R.O. Quebec, Dec. 4, 1830.
38 W.O.44/20, p. 235, By to Couper, May 19, 1830.

Act. Only the governor-in-chief had the authority to purchase because the canal was technically a Colonial Office project. By had exceeded his powers and disobeyed orders; Ellicombe wanted an explanation.[39] After an investigation, however, the Ordnance concluded it was impossible to determine exactly how and by whom the land should have been acquired because two departments of state had been involved.[40] The Treasury later concluded that By had acted unwisely but no clear explanation of what had gone wrong was provided.

The first and least troublesome consequence of By's land arrangements after the Ordnance took control of the waterway involved the settlement of damage claims stemming from canal construction. Writing in England in December 1833 By assured the Ordnance that "Government *will have very little to pay*" in damages because of the precautions he had taken. As evidence he cited a most reasonable settlement he had concluded with Nicholas Sparks of Bytown.[41] Sparks—as will be shown—turned out to be the most difficult antagonist the Ordnance encountered. In any case, Colonel Nicolls had already written from Canada that much money would be needed for damage claim settlements.[42] So difficult was the claims problem that the Ordnance, unable to solve it through normal procedures, had to appoint special arbitrators. The first two men made little progress;[43] the third, George Adams, deputy assistant commissary general, worked hard from 1835 to 1846. Initially it was impossible to determine what the final damage costs would be and the Ordnance asked the Treasury for authority to pay out whatever was needed. This request was refused, so in 1834 the Ordnance asked for £10,000. The next year it asked for £20,000.[44] In February 1837 Seth Thomas complained that the arbitrator was proceeding too slowly. Only £2,600 in claims had been settled to that time.[45] Seven years later Elliott reported that £21,998 had been paid out for damages and court costs since 1832.[46] Adams took three more years before closing the accounts, but a few claims still remained.[47]

Of these, the biggest was that of Nicholas Sparks. This man, an original settler of what became Bytown, was an Irishman who by the early 1820s owned several hundred acres straddling the northern terminus of the future Rideau Canal. As the community grew under the stimulus of the canal he became a successful and unscrupulous land speculator.[48] In 1854 he was still demanding that the Ordnance return some one-hundred acres of Bytown property to him. Colonel By, he insisted, had robbed him by means of false pretenses. In 1826 he had agreed to sell By sufficient land for canal

---

39   Ibid., p. 242, Minute by Ellicombe, Sept. 1, 1830.
40   Ibid., p. 246, Ordnance minute, Sept. 8, 1830.
41   W.O.44/21, pp. 356-57, By to Fanshawe, Dec. 16, 1833.
42   Ibid., pp. 30-31, Nicolls to Pilkington, Oct. 25, 1833.
43   C.O.42/257, p. 269, The R.O. Quebec to Major Airey, June 29, 1835.
44   W.O.44/23, pp. 177-80, Ordnance minute, 1837.
45   W.O.46/135, pp. 107-09, Minute by Thomas, Feb. 14, 1837.
46   W.O.44/45, p. 119, Elliott to Byham, July 25, 1843.
47   Ibid., p. 224, G. Adams to the R.O. Quebec, Apr. 18, 1846.
48   R.B. Sneyd, "The Role of the Rideau Waterway," p. 173.

purposes. Subsequently By had taken more on the pretext of military necessity and by authority of the Rideau Canal Act, but against Sparks' will. The additional property had not in fact been used for the canal and therefore By's action was illegal. This charge was perhaps tenable in law, but the Ordnance had actually paid for the extra land in question. Even so, Sparks regained much of what he claimed through a special act of the Canadian legislature in 1846. The remaining disputed property was in the heart of a thriving town, and in 1854 he still wanted it back on the grounds that the Ordnance had acted illegally, arbitrarily and greedily.[49]

Elliott, who battled Sparks in the 1840s and almost failed to acquire legal title to all Ordnance properties as a result, summarized the department's position on this claim in 1855. Sparks was wrong about the alleged misdeeds of Ordnance officers, had been treated more than fairly, and had regained more land than he deserved. In Elliott's opinion the case was closed. It had been "the subject of unceasing correspondence and vexatious litigation."[50] The same could be said for the whole of the Rideau damage claims problem.

More critical still was the simultaneous Ordnance struggle to acquire a Canadian vesting act, not successfully concluded until 1843. In 1828 Upper Canadian legal authorities declared the imperial Ordnance Vesting Act of 1821, giving the department title to all military property in the empire, inoperative in the province. The London authorities became fully aware of this problem only with the building of the Rideau Canal.

The attorney general's office of Upper Canada followed its 1828 ruling by applying the Rideau Canal Act in a way which completely denied Ordnance rights to Canadian lands. All land titles acquired under this act were conveyed to the Crown, to Lord Dalhousie prior to 1827, or to Colonel By thereafter. In 1832 the Crown's legal advisers upheld the Upper Canadian ruling. They recommended that the 1821 Vesting Act should be amended so it would apply to Canada and other colonies. Parliament passed the amendment, but the act still proved inoperative outside Great Britain. The only remedy was a provincial vesting act, and James Smith, solicitor to the Ordnance, asked the Upper Canadian attorney general to acquire such an act in August 1834. The Canadians did nothing; in 1836 some Rideau lands were still vested in Dalhousie and the remainder in the heirs of John By. James Smith reported that Dalhousie, still living, could easily transfer his title to the Ordnance, but the heirs of By were hostile and would create difficulties.[51]

In Canada Ordnance officers already had other troubles associated with land titles. Captain Bolton found he could not collect rents from local inhabitants who had entered into leases with By in 1830; arrears amounted to £1,676 Halifax currency by 1835.[52]

49   W.O.44/16, pp. 616-19, Sparks to the Ordnance, Sept. 29, 1854.
50   Ibid., pp. 620-27, Elliott to the R.O. Montreal, Jan. 8, 1855.
51   W.O.44/28, pp. 38-42, Minute by James Smith, Feb. 1836.
52   W.O.55/284, pp. 1-52, Eaton and Elliott report, Nov. 9, 1840.

When Bolton twice took the department's debtors to court his suits were rejected on the grounds that the Ordnance held no land in the province. The department appealed for help with a vesting act to the Colonial Office but the colonial secretary and his governors had more pressing Canadian political problems.[53] In 1836 the officers from Quebec reported that neither of the two provincial legislatures showed a willingness to pass the required legislation. The Lower Canadians had given a vesting bill a reluctant second reading before the ending of the session terminated the process of law-making.[54] The respective officers tried to introduce their bill again in 1837 and 1838 without success.[55] The political crises surrounding the rebellions of 1837 were absorbing all the attention of the provincial legislators.

The aftermath of the rebellions, however, did give the Ordnance temporary title to its Lower Canadian lands. While the Lower Canadian legislature was suspended in 1839, governor general Charles Poulett Thomson promulgated an Ordnance vesting ordinance effective until the united parliament of Canada was constituted.[56] But Upper Canada's legislature was not suspended. During the sessions of 1839 and 1840 the majority in the assembly expressed the opinion that, first, military lands vested in the Crown could be better defended in courts of law than lands vested in the respective officers and, second, reserves not immediately required for defence works would be of more use to the province than the Ordnance.[57]

When the Ordnance commissioners reported these details, chief clerk Seth Thomas lost his temper, attacking Canadians in general and the inhabitants of Bytown in particular. In a confidential memorandum to Byham, the master general's secretary, Thomas explained that the Bytown residents were petitioning the provincial government to convert their Ordnance leasehold into freehold on the grounds that the department's tenants did not have the franchise. Thomas felt the franchise problem was their own business; they could solve it some other way. If the Ordnance was to carry out its duties in Canada it had to raise all the revenues possible out of the Rideau holdings.[58] The chief clerk complained that

> The case sought to be established by the Colonists, in all its bearings appears to be altogether too unreasonable to be entertained. The facts of the case are simply these. The British Government wish to construct a Canal, which, at the same time that it shall add to the Military Protection, shall, in other respects be greatly beneficial to the Colony—the mother country furnish Funds exceed-

---

53 W.O.47/1690, Ordnance minute, Aug. 19, 1835, and C.O.42/428, p. 67, Byham to Hay, Aug. 19, 1835.

54 W.O.44/28, p. 21, The R.O. Quebec to Byham, July 6, 1836, p. 18, Smith to Byham, Aug. 27, 1836, and p. 10, Smith to Byham, Oct. 1, 1836.

55 W.O.55/284, pp. 1-52, Eaton and Elliott report, Nov. 9, 1840.

56 W.O.44/29, pp. 13-14, Copy of ordinance 2 Vic: Cap. XXI, 1839, vesting the R.O. in Lower Canada with all military lands.

57 W.O.55/284, pp. 1-52, Eaton and Elliott report, Nov. 9, 1840.

58 W.O.44/23, pp. 8-22, Thomas to Byham, Jan. 30, 1841, and W.O.1/537, pp. 460-64, Byham to Trevelyan, Mar. 27, 1841.

ing a Million Sterling—and under authority derived from an Act passed in the Legislature of the Colonists themselves, the Crown absolutely purchases, out of monies it has provided, the fee simple of the Lands in question with the proviso of deputing the Management or Proprietorship of these lands to whichsoever of its Officers it may in its wisdom deem proper. In due time this Proprietorship is transferred to the Principal Officers of the Ordnance and when they seek to exercise the rights of Ownership—they are met at every turn by questions as to the validity of the Crown's title, or that of the Department deputed by the Crown, which in point of fact is the same thing. The object of this opposition and these difficulties being, to establish a case to recover back the Lands that have been purchased; under the denomination of a change of tenure: and that too notwithstanding the vastly increased value of the adjoining Lands by the construction of the Canal which must have been highly beneficial to the neighbouring Proprietors.[59]

Thomas then turned to individuals. William Draper, provincial attorney general and member for Russell County—in which Bytown was partly situated—had advised Ordnance tenants not to pay rent. He and Stewart Derbishire, member for Bytown, had pledged themselves to resist all Ordnance efforts to get legal title to the Rideau lands and were now campaigning in the united parliament to this end. It was rumoured that governor general Lord Sydenham was sympathetic. There was still more perfidy in Canada:

The result of such advice has been that the bulk of the Tenants refuse payment of Rent, and one of them, Daniel Johnson, who has been a nuisance to the Dept. for the last 7 Years, and should long since have been ejected, not only retains Land for which he has never paid any Rents, but by collecting a Mob, and by intimidation, prevents a well disposed settler from occupying certain lots let to him by the R.O.; and by this means, having himself obtained forcible possession of these Lots, he actually sells them to another individual for the sum of £100, although it is clearly shewn he has no title or claim to the Lots in question.[60]

As for the last attempt to get a vesting act through the Canadian legislature, Thomas accused justice Christopher Hagerman of deliberately suppressing the Ordnance bill while pretending he was trying to help. The chief clerk therefore proposed that if a Canadian vesting act could not be procured soon, Ordnance title should be forced on the colonists by means of a royal charter.

The master general and board accepted these views and acted on them.[61] Richard Byham sent a summary of the Thomas indictment to the Treasury, requesting a royal vesting charter if a provincial act could not be obtained.[62] The Treasury in turn passed

---

59 W.O.44/23, pp. 8-22, Thomas to Byham, Jan. 30, 1841.
60 W.O.46/136, pp. 161-84, Minute by Thomas, Dec. 8, 1841.
61 W.O.44/24, p. 449, Minute by the master general, Feb. 25, 1841.
62 W.O.1/537, pp. 460-64, Byham to Trevelyan, Mar. 27, 1841, and W.O.1/538, pp. 534-36, Byham to Trevelyan, Dec. 31, 1841.

these papers to the Colonial Office.[63] With Treasury and Colonial Office support, Ordnance commissioner Elliott took charge of the effort to push through the Canadian act.

His task was most difficult. In 1841 the Ordnance was seeking vesting acts throughout the empire,[64] but in Canada the crisis over legal title revolved around the Rideau properties and the difficulties were greater. Furthermore, the new governor general, Charles Bagot, was hostile to Elliott's mission.

On receipt of a summary of the Thomas indictment, Bagot denied all Ordnance accusations against Canadian members of the legislature, claiming the allegations against them to be based on irresponsible "electioneering rumors and misrepresentations of the day." Turning to Elliott and Eaton, he stated to the colonial secretary that

> . . . your Lordship will I am sure feel the extreme course adopted in this instance by the Ordnance Commissioners, and the inconvenience which must be caused to the public service by a repetition of it. The Ordnance Commissioners are here in a very anomalous position. Unrecognized by either the Civil or Military Authorities of the Province and under the control of neither, they appear to consider it their duty to criticize the conduct of every Department of the Government in their Report to the Master General and Board of Ordnance. . . . If any ground exists for impugning the conduct of a public Officer, I respectfully submit that it is to myself, and not to a subordinate Department at Home that complaint should be made. . . .[65]

Like other governors, Bagot resented Ordnance interference which undermined his authority and complicated his political tasks.

Nevertheless, Elliott persevered in his quest. He presented a vesting bill to the legislature in 1842, losing it because of the shortness of the session.[66] The next year he tried again, this time in the face of a resistance movement organized by the allies of Nicholas Sparks. When Elliott presented his bill to the assembly at Kingston, Sparks and 145 inhabitants of Bytown petitioned the legislature for a restoration of all their lands. The house of assembly convened a special committee to consider the petition. The chairman was William Stewart, member for Bytown, and he introduced a bill which would enable claimants of Rideau lands to seize Ordnance property merely on the strength of their word. Elliott and the respective officers protested to the governor general.[67]

In the meantime, Captain Edward Boxer of the navy, then working with Ordnance officers on a Canadian defence survey, and a legislative councillor allied to the Sparks faction, the honourable Thomas McKay, approached Elliott with a proposal for a private agreement between Sparks and the Ordnance. Elliott rejected this move at first, but the

---

63   W.O.1/537, pp. 458-59, Trevelyan to Stephen, Apr. 22, 1841.
64   W.O.1/538, p. 514, Stanley to Bagot, Oct. 5, 1842. Stanley enclosed copies of a despatch being sent to all governors, directing them to obtain colonial Ordnance vesting acts as soon as possible.
65   Ibid., pp. 44-47, Bagot to Stanley, Mar. 26, 1842.
66   Ibid., pp. 511-12, Elliott to Byham, Sept. 26, 1842.
67   W.O.44/35, pp. 189-92, Elliott to Byham, Dec. 9, 1843. See also *Provincial Statutes of Canada*, 7 Vic. Cap: 11, 9 Dec., 1843.

Stewart committee ruled that the house would only accept a vesting bill which restored the Sparks property to its claimant. Elliott gave in, and with the aid of Gustavus William Wicksteed, law clerk of the assembly, steered the amended bill through in December 1843. Although regretting this amendment, he still expressed satisfaction with the act. It solved a problem which had plagued the department for over a decade; it also destroyed a potentially dangerous legal argument then being advocated by Stewart, that the Act of Union joining the provinces in 1840 gave the new provincial government control of all unused military lands.[68]

Yet the vesting act battle was not over. The Ordnance legal advisers discovered that the Sparks amendment was so badly worded it was inoperative in law. Sparks insisted he had regained title to some eighty-eight acres and proceeded—along with Stewart and others—to take possession. The respective officers sued for trespass and won. Therefore Sparks again petitioned the legislature in 1845, charging the Ordnance with trickery and breach of faith. Stewart once more became chairman of a special committee to consider the petition; he was himself one of the trespassers sued by the department. When the committee attempted to give Sparks the eight-eight acres by means of a special bill the respective officers had the bill reserved by the governor general.[69] In their view the Ordnance could hardly be blamed for defects in the Sparks amendment of 1843 and in any case the amendment had been forced on the department unfairly. They were attempting to safeguard the public interest against the questionable demands of self-seeking individuals.[70] Hopefully the whole dispute, "which has proved one of unceasing vexation to the Department for the last Seventeen years," could finally be ended.[71]

But the Colonial Office forced the Ordnance to yield on the Sparks issue. The 1843 amendment, stated A. W. Hope, had been intended to function effectively by all parties, and Lord Stanley did not feel the imperial government could justifiably disallow the 1845 bill.[72] So the conflict over the vesting act was finally over, although there were further troubles with Sparks and others. According to the commanding Royal Engineer in Canada, Colonel W. E. C. Holloway, in 1844 there was still continuing confusion and litigation over departmental land boundaries.[73] All the same, the Ordnance was now finally recognized as a legal property holder in Canada.

* * * * *

68   Other colonial Ordnance vesting acts were obtained throughout the empire in the late 1830s and early 1840s. For example, New South Wales passed such an act in 1840. See W.O.47/1969, p. 6192, Ordnance minute, May 12, 1843.

69   W.O.44/35, pp. 273-78, Holloway to Mulcaster, Apr. 25, 1845. See also ibid., pp. 349-53, Report of the special committee of the Canadian House of Assembly on the petition of Sparks, 1845.

70   W.O.44/30, pp. 21-29, The R.O. Montreal to the Ordnance, Apr. 23, 1845, and pp. 17-20, Elliott to J. M. Higginson, Nov. 25, 1844.

71   Ibid., pp. 21-29, The R.O. Montreal to the Ordnance, Apr. 23, 1845.

72   W.O.44/35, A. W. Hope to Byham, Aug. 23, 1845. See also *Provincial Statues of Canada*, 9 Vic. Cap: 42, 1846.

73   W.O.44/30, p. 52, Holloway to the inspector general of fortifications, Oct. 11, 1844.

The trouble was that it had taken too long to get the vesting act. By 1845 the chances of raising substantial funds from property were slipping away from the department. In particular the Ottawa-Rideau system, until the mid-1840s the largest potential asset in Canada, was now becoming a forlorn hope.

The military waterway remained incomplete. After 1832 Ordnance officers did their best to expand all of it to the thirty-three foot wide scale of the Rideau Canal, thereby increasing its military value and also revenues from civilian traffic. In 1836 Captain Bolton argued that if the three Grenville Canal locks completed on the twenty-foot scale prior to By's enlargement of the project could be widened, and the Lachine Canal bypassed by a wider alternative at St. Ann's, the income from traffic would double.[74] His was one of many attempts to renew the work stopped in 1834. The local inhabitants sent numerous petitions to the Ordnance and other imperial authorities to try to improve the communication system on which the growth of their communities depended.[75] In these circumstances the Ordnance and Treasury suggested that the Canadians should pay for the additional works. In March 1836 the governor-in-chief proposed this solution to the Lower Canadian assembly,[76] but the colonists were more interested in raising money for canals on the St. Lawrence. In 1842 and 1843 Seth Thomas again renewed the enlargement campaign; Sir Frederick Mulcaster, the inspector general, explained that the necessary expenditure could not be justified.[77] The final burst of effort came in 1845 and 1846 during the Oregon crisis between Britain and the United States. Governor General Sir Charles Metcalfe, supported by the Duke of Wellington, temporarily gained Treasury sanction for enlargement but the dispute with the United States was settled and no action was taken.[78]

By this time the Ordnance was thoroughly frustrated with the Ottawa-Rideau waterway and contemplating the transfer of the system to the Canadians. Colonel By had predicted the canals would bring rich rewards, persuading the Treasury to hold on to them under Ordnance auspices; instead they remained a continual burden. In 1830 By had asserted that Rideau Canal tolls alone would produce over £40,000 a year in revenue. In February 1832 he had raised the estimate to about £50,000 per annum.[79] As usual, he had been too optimistic.

---

74 W.O.44/22, pp. 61-70, Bolton to the R.O. Quebec, Feb. 20, 1836.

75 C.O.42/429, pp. 96-110, Colborne to Glenelg, Feb. 29, 1836. See also Sneyd, "The Role of the Rideau Waterway," pp. 71-72, and p. 132.

76 W.O.44/22, pp. 54-56, The R.O. Quebec to Byham, Mar. 4, 1836, and W.O.44/16, pp. 300-02, Thomas to Byham, Nov. 1842.

77 W.O.44/16, pp. 303-04, Mulcaster to Byham, Dec. 9, 1842, and W.O.44/45, pp. 111-13, Thomas to Byham, Oct. 4, 1843.

78 W.O.1/552, pp. 29-33, Metcalfe to Stanley, Feb. 1845, and W.O.1/555, pp. 139-50, Wellington to Gladstone, Apr. 29, 1846. See also Sneyd, "The Role of the Rideau Waterway," pp. 132-33.

79 W.O.44/22, pp. 171-74, Thomas to Byham, May 11, 1835.

Canadians using the canals had a vital interest in the rate of tolls. They met every Ordnance attempt to increase revenue with firm and effective resistance.[80] By 1835 the average annual Rideau revenues amounted only to some £3,059. At this time Seth Thomas still felt that the predicted income of over £40,000 might be realized with improved management.[81] On his suggestion the Ordnance commissioned Seth Thomas Junior as investigator. After examining canal management techniques in Britain and the United States the younger Thomas was to join Charles Lennox Rudyerd, Ordnance paymaster on the Rideau, and act as adviser.[82]

In the meantime Rudyerd and Captain Bolton were protesting that no more revenue could possibly be squeezed out of the canal; By's predictions had been totally wrong.[83] The details Bolton presented in support of his arguments demonstrated the problems of revenue raising. By, the Captain stated, had made his original calculations on the basis of a scale of tolls he had drawn up in November 1830. In 1832, responding to Canadian complaints, he had reduced these tolls to such an extent that according to his own 1830 calculations the revenue would have declined to £16,000 per annum. Some 8,000 cabin passengers, bringing in £2,000 a year, and 20,000 tons of goods at an annual rate of £20,000 had been expected by the Colonel; apparently he had assumed that all water traffic between Montreal and Lake Ontario would employ his route. Bolton, however, pointed out that while the Rideau Canal was carrying much upstream traffic and providing invaluable service to Canadian commerce, the downstream traffic from the Great Lakes to Montreal used the St. Lawrence despite rapids. The St. Lawrence route was fifty miles shorter. Thus, in 1834 only 825 passengers and 3,611 tons of goods—at 7/6 a ton— had passed through the canal. Bolton also warned that when the Canadians finished their recently commenced St. Lawrence canal system toll revenues on the military canals would decline further.[84] Little could be done to improve matters by changed methods of management. The younger Thomas joined Rudyerd with his accumulated knowledge of British and American canals in January 1836 but there was no subsequent improvement in Rideau finances.[85]

Until 1842 income from tolls did not even cover maintenance costs. In the first six months of 1836 the Rideau Canal brought in some £644 while running costs came to £4,382.[86] By contrast the situation on the Ottawa canals was better. Between 1835 and

80 W.O.44/20, pp. 102-05, Petition of the citizens of Kingston to Colborne, May 1, 1832, R.G.8, C 58, pp. 61-68, Petition of Porter Grimmell and Co. and 450 other individuals to Colborne, Apr. 1834, and W.O.44/25, pp. 241-42, Stephen to Byham, June 11, 1842. For a detailed account of the conflict between Canadians and the Ordnance over tolls from 1832 and 1845 see Sneyd, "The Role of the Rideau Waterway," pp. 44-53, 97-118, 122, 128 and 130.
81 W.O.44/22, pp. 171-74, Thomas to Byham, May 11, 1835.
82 Ibid., pp. 75-94, Seth Thomas Junior to Rudyerd, Jan. 12, 1836.
83 Ibid., p. 145, Rudyerd and Bolton to the R.O. Quebec, Aug. 24, 1835.
84 Ibid., pp. 61-70, Bolton to the R.O. Quebec, Feb. 20, 1836.
85 Ibid., pp. 75-94, Seth Thomas Junior to Rudyerd, Jan. 12, 1836.
86 Ibid., p. 45, Half-yearly Rideau Canal accounts, by Rudyerd, June 30, 1837.

1836 Rideau revenues declined by £3,005; Ottawa revenues rose by £382.[87] Early in 1840 the respective officers suggested that the Rideau Canal should be given to the Canadians to stop this annual drain. Still, Seth Thomas argued that if the Ottawa canals could produce a profit the Rideau waterway might eventually become profitable as well.[88] In 1841 he reported that for all the Ordnance canals the annual income was now £10,500 and the outlay £12,058, leaving a deficit of only £1,558.[89] By this time, however, the Canadians were beginning to make good progress on the St. Lawrence project, and Thomas was sufficiently concerned to suggest that ". . . it may be deserving the serious consideration of the Master General and Board, and of her Majesty's Government, how far it may be judicious to raise up a rival interest to a Work which has cost this country so vast a Sum, and at the risk of the failure of both in a commercial view, when there already exists an excellent water communication, sufficient for all the purposes of commerce."[90] He wanted the St. Lawrence project stopped because it threatened not only his department's income but the military security of the province. The unpopular and isolated Ordnance of course had no hope of forcing so drastic an action.

At the eleventh hour, in 1842, the Ordnance canals finally showed a modest profit of £852. For the next three years this figure rose steadily, reaching a peak of £2,731 in 1845.[91] In 1846 there was once more a loss of £2,490.[92] The St. Lawrence canals, while not completely finished, were partly opened for business in 1845 and had the expected effect on Ordnance canal revenues. Thomas lamented that "the Tolls on the St. Lawrence, are so much lower than those at present on the Rideau and Ottawa, that the results must be inevitably prejudicial to the Ordnance. . . ."[93] From 1845 the idea of transferring the military canals to the Canadians gained increasing departmental support. Along with the working of hostile forces in London, the failure to make colonial possessions pay was the beginning of the end for the Ordnance in Canada.

The original transfer of military property to the department, and especially the acquisition of Colonel By's waterway, had thus raised hopes among Ordnance officers that they might yet finish a good portion of Wellington's fortification scheme without Treasury support. But the Canadians and their governors opposed the Ordnance, regarding it as an alien meddler in colonial affairs. They had no sympathy for an organization seeking to provide for their security at no expense to Britain.

---

87   Ibid., pp. 11-13, Thomas to Byham, Mar. 30, 1837.
88   W.O.44/16, pp. 257-58, Thomas to Byham, Apr. 10, 1840.
89   Ibid., pp. 268-70, Thomas to Byham, Jan. 7, 1841.
90   W.O.44/49, pp. 34-36, Memorandum by Thomas, Jan. 30, 1841.
91   W.O.44/16, pp. 300-02, Thomas to Byham, Nov. 1842.
92   W.O.44/49, p. 529, Accounts of receipts and expenditures on the Ordnance canals, 1844 to 1848, by the R.O. Montreal, Mar. 21, 1848.
93   W.O.44/25, pp. 44-45, Thomas to Byham, Aug. 7, 1845.

Chapter IX

Dissolution

During its last decade in the province the Ordnance was under continuous assault from both colonists and imperial reformers; nevertheless, its already frustrated officers sought to carry on to the end. At least they could try to preserve those colonial properties which might still produce revenues for military works, while attempting to liquidate liabilities like the Ottawa-Rideau system. But the Canadians would only accept responsibility for the waterway if profitable military lands were included in the transfer. In other words, to rid itself of a burden the department was pressed to surrender its assets. To the newly self-governing colonists the Ordnance appeared to have no valid role to play in their community. Nor indeed did it have a place left in an imperial administration bent on centralizing control of all military departments under the cabinet.

After 1846 Earl Grey and his fellow reformers began a piecemeal dissolution of the Ordnance, starting in the most politically independent colonies. They sought to reduce Britain's overseas military expenses on the principle that self-government entailed greater responsibility for self-defence, and their first major target was the unpopular Ordnance. Like the Canadians, the reformers sought to divest the Ordnance of its land, enabling the formation of cheap interim colonial defence forces which would ease the anxiety of overseas subjects unwilling to pay for their own security. This land, if promised to the Canadians, might also induce them to accept the task of keeping fortifications, barracks and military canals in good order. The department was too weak to withstand this combination of pressures.

Although individual Ordnance officers continued to urge retention of the canals, the department adopted the idea of transfer late in 1844. Master General Sir George Murray took the initiative and argued strongly that although the canals had been built for military purposes, in practice they served predominantly civilian interests. Civilians, therefore, should pay for the upkeep. The Canadians were constantly complaining about defects in the waterway and had petitioned the imperial government for enlargement of the small-scale locks; if they needed such improvements they should provide the funds. It would be best for everyone if they took over the waterway, on condition only that it was kept in a proper state for military and naval use in time of war. Murray concluded by warning that,

> So long as the present System continues, of this line of Navigation, being in the great part of its extent under the Management and control of Military Authorities, and with a *door open* for claims of an *undefined* amount upon the resources of the Mother Country, so long will the navigation itself remain (I apprehend) upon a precarious footing; and one likely to give occasion to difference of opinion, and to disputes of difficult adjustment.[1]

Murray's proposal was adopted but the Ordnance insisted that the conditions of transfer had to guarantee the department's Canadian interests. As well as granting free passage for military and naval transport, the province should pay at least half the costs of completing and improving the waterway. This was thought only fair, since the canals, built at the cost of over a million pounds, had already done much to foster the growth of the commerce of the colonists, who were clearly the chief beneficiaries of the transfer.[2] The chancellor of the exchequer, Henry Goulburn, agreed with these Ordnance proposals and pledged Treasury support.[3]

Unfortunately the Canadians were not interested. Governor General Sir Charles Metcalfe explained in 1845 that the provincial government had no intention of assuming the expenses of canal maintenance.[4] The following year Earl Cathcart, Metcalfe's successor, reported the same attitude. But he was a governor for once sympathetic to the Ordnance. In his opinion,

> . . . litigious proceedings, charges, and angry recriminations of the Community bordering on, and using the Waters of this Navigation, were unceasing and vexatious; that the disputes, Lawsuits, and consequent expenses, which the Ordnance have had to encounter, are, and apparently will be, almost endless; that the obliquy, to which the Department is exposed, by the adverse strictures of a Party Press, in consequence of the great extent of the Ordnance Lands uncultivated, occasions this Branch of the Imperial Government to be held in much unpopularity. . . .[5]

Cathcart concluded after discussions with Colonel Holloway of the engineers that the best counter to provincial disinterest was to lease the canals to the shipping companies currently using them.

After the Ordnance and Treasury agreed that this idea should be pursued further,[6] the Lords Commissioners requested in July 1847 that the Colonial Office instruct the governor general to convene an Ordnance canal investigating commission.[7] Delays

---

1 W.O.1/553, pp. 505-06, Minute by Murray, Nov. 16, 1844.
2 W.O.44/49, pp. 9-11, Ordnance memorandum, Dec. 16, 1844.
3 W.O.1/553, pp. 513-14, Goulburn to Stanley, Jan. 14, 1845.
4 W.O.1/552, pp. 73-74, Metcalfe to Stanley, Mar. 15, 1845.
5 W.O.44/49, pp. 299-305, Holloway to Burgoyne, Aug. 12, 1846.
6 W.O.1/558, pp. 401-02, Minute by the master general, Sept. 14, 1846, and W.O.44/49, p. 325, Treasury minute, Feb. 19, 1847.
7 W.O.1/558, pp. 451-52, Byham to Stephen, July 9, 1847.

caused by the usual problems of interdepartmental communication, however, held up proceedings until the following year. By then Cathcart had been replaced by Lord Elgin and the commission did not even give Cathcart's plan serious consideration. The commissioners, mostly Ordnance officers, reverted to the Murray transfer idea; Elgin explained to them that the Canadians, burdened with their own public works commitments, still did not want the canals.[8] Thus the commission ultimately recommended retention of the waterway by the department.

The commission's report stated that the canals were too important for Canada's military safety to be entrusted to a local government which did not want them in any case. Neither the province nor private companies would keep them in a proper state of repair. It was thought that there might still be a chance of making them pay by better management, revised tolls and a much more thorough exploitation of rents from adjacent Ordnance lands.[9] Neither the Treasury nor Earl Grey at the Colonial Office accepted these proposals; both insisted on transfer, with the result that there was interdepartmental deadlock.[10] Ordnance hopes for revenues were again dashed and the Canadians refused to accept transfer without the inclusion of profitable lands the department wanted to keep. Despite persistent efforts by the Treasury and the Colonial Office it was impossible to initiate direct transfer negotiations until 1853.[11]

The Colonial Office exerted particularly strong pressure on the Ordnance. In 1846 Earl Grey began his campaign to improve Britain's defences and to raise the quality of the armed forces. Parliament had refused to expand the military budget to an adequate level to meet what Grey, Wellington and others regarded as a home defence crisis brought on by shifts in the military balance of power in Europe. Therefore the secretary of state for war and the colonies decided it was necessary to concentrate the army in the British Isles and to start a withdrawal of the legions from the colonies.[12] Grey knew the colonists would resist, for they still considered that the imperial government had a duty to provide them with military protection. Thus he approached the self-governing colonies first, arguing that increased political independence entailed assumption of increased responsibility for local defence. As he expressed it to Lord Elgin in 1848, "In N. America the only charge of any moment upon this Country is that of the Military defence of our dominions there, an expense wh. I am sure admits of large reduction if we can but carry the Colony with us. They certainly ought to bear a part of the charge of

8   W.O.44/49, p. 528, The R.O. Quebec to the governor general's principal secretary, Mar. 24, 1848, and p. 529, Sullivan to the R.O. Quebec, July 3, 1848.

9   Ibid., pp. 536-56, Report of the Ordnance canal commission of 1849, Feb. 28, 1849.

10 W.O.1/561, pp. 296-97, Trevelyan to Merivale, Feb. 14, 1849 and R.G.1, E4, Vol. 5, pt. 1, p. 143, Grey to Elgin, Feb. 20, 1849.

11 W.O.44/49, p. 253, Bruce to the R.O. Montreal, Apr. 3, 1850, and W.O.1/564, pp. 1-19, Report of a Treasury committee on colonial defence expenditures, July 22, 1851.

12 C.O.885/1, Confidential memorandum on the state of the army, by Earl Grey, Oct. 17, 1846. See also Bourne, *Britain and the Balance of Power*, pp. 172-75; Hitsman, *Safeguarding Canada*, pp. 150-60; and Schuyler, "The Recall of the Legions."

their own protection now that they are allowed such complete self Govnt. . . ."[13] To "carry the Colony" he concentrated first on removing the unpopular Ordnance.

Grey's attack took the form of a plan to transfer the department's possessions to the colonists. He did not distinguish between assets like profitable lands or liabilities like the Canadian canals; everything should be given up. In his words,

> There is nothing wh. is now managed in a manner So unsatisfactory and leading I am persuaded t. so Much extravagance as this part of our service partly from the constitution of the Board of Ordnance, but more from the utter impossibility of such business being managed efficiently and economically from a distance. In Australia I am refusing to undertake any works unless the Colonies will pay for them and in N. America it wd. I am persuaded be an immense advantage if the Ordnance cd. make over their property to the Provincial Govmts the latter undertaking to provide in future Barracks for the Troops and to make all necessary fortifications.[14]

Grey decided to start with New South Wales and South Australia, which were less likely than Canada to resist the imposition of the burden of Ordnance costs.

With the Treasury behind him the colonial secretary ordered governor Sir Charles Fitzroy of New South Wales to execute transfer by 1850. From that date the colonists were to take over all barracks and fortifications and to keep them in proper repair for use by British troops in emergencies. Grey imposed these conditions without consulting the government of New South Wales or the Ordnance.[15] Both acquiesced and he felt he had made a good beginning; as he stated, "N. S. Wales kicked a good deal at first but the people there are beginning to understand that they cannot have the advantages without the burdens of self govnt—S. Australia has agreed without demur to undertake the Barracks."[16] The surrender of the Australians would be a good example with which to press the Canadians. On March 14, 1851, Grey therefore informed Elgin that it was now time to apply the Australian transfer arrangement to Canada.[17]

But the Canadians would not be coerced. They refused to assume expensive responsibilities without proportionate rewards. This gave the Ordnance time for a rear-guard action in defence of its Canadian lands.

Nevertheless, in February 1853, the Treasury, at the end of its patience, informed the Colonial Office that after September 30, 1854, it would pay no more military canal expenses. The respective officers would have to surrender the waterway at that time. A decision about Ordnance estates bordering the canals would be made at the completion of negotiations for the transfer of all departmental properties.[18]

---

13   Doughty, ed., *The Elgin-Grey Papers*, p. 186, Grey to Elgin, July 7, 1848.
14   Ibid. See also p. 249, Grey to Elgin, Nov. 10, 1848, and p. 259, Grey to Elgin, Dec. 1, 1848.
15   W.O.1/521, pp. 159-60, Minute by Grey, May 28, 1849, pp. 237-46, Grey to Fitzroy, Nov. 21, 1849, and p. 307, Trevelyan to Merivale, Nov. 1, 1849.
16   Doughty, p. 721, Grey to Elgin, Oct. 25, 1850.
17   Ibid., pp. 1584-89, Grey to Elgin, Mar. 4, 1851.
18   W.O.1/567, pp. 156-57, Trevelyan to Merivale, Feb. 28, 1853.

The Canadian executive council arranged to assume canal management after the deadline, but stipulated that certain conditions would have to be met before action was taken. Canal maintenance costs were declining, from £18,744 in 1848 to £10,563 in 1852, but revenues were also declining, from £5,010 in 1848 to £2,826 in 1852. On the other hand, in the same period annual rents from adjacent Ordnance lands had risen from £1,083 to £1,809. Therefore the executive councillors decided, first, that they would investigate the possibility of lowering canal costs and raising revenues and, second, that the imperial government should be asked to include the military lands in the transfer agreement. If the canals went to the province the lands would be of no further use to the Ordnance; they would, however, act as a strong inducement impelling the Canadians to maintain the waterway efficiently.[19] In short, Britain would have to pay a price in Ordnance lands to get rid of the canals.

Meanwhile the imperial government decided that the canal transfer terms should be arranged by consultation between the Treasury, the Colonial Office and the Ordnance. This gave the latter a chance to exert some pressure. The master general drew up what his department considered minimum conditions. First, the province would have to keep up the canals as efficiently as the Ordnance, second, all military and naval transport would have to have free passage, and, third, there would have to be a provincial guarantee of all existing Ordnance land titles. Fourth, and most important, ". . . if it be decided that the lands not absolutely necessary for the maintenance of the Canals from whatever source obtained should remain at the disposal of the Ordnance, all such parts of them as are not necessary for Military purposes should be sold from time to time, and the proceeds should constitute a fund for military works in the colony. . . ."[20] So even at this late date the Ordnance hoped to use its Canadian possessions to finance defence works.

These four conditions once more threatened the transfer arrangements with deadlock. The Ordnance terms, communicated by the Treasury, were unacceptable to the Canadians. They wanted full freedom to determine the efficiency of canal maintenance themselves; the council doubted if maintenance costs could be met without Ordnance land revenues. The inclusion of the land in the transfer arrangements would go far to influence the provincial parliament to assume responsibility for such expensive military works.[21] Of course the Ordnance was unmoved by these arguments. To the department there was no good reason why the imperial government should surrender an income of £2,000 a year.[22] To resolve this impasse the Treasury resorted to coercion and overruled the Ordnance. Sir Charles Trevelyan explained that,

> My Lords are however disposed to look with favour upon any proposed concession which though not demanded upon them upon principles of strict right, nor perhaps, if judged by the particular case, even of equity, yet may be of a nature to aid and further the prosecution of what My Lords understand to

19 W.O.1/566, pp. 78-80, Report of a committee of the Canadian executive council, May 13, 1853.

20 W.O.1/567, pp. 109-10, Butler to Merivale, Jan. 23, 1853.

21 W.O.1/566, pp. 148-49, Report of the Canadian executive council, Sept. 14, 1853.

22 W.O.1/567, pp. 267-68, Butler to the Treasury, Oct. 28, 1853.

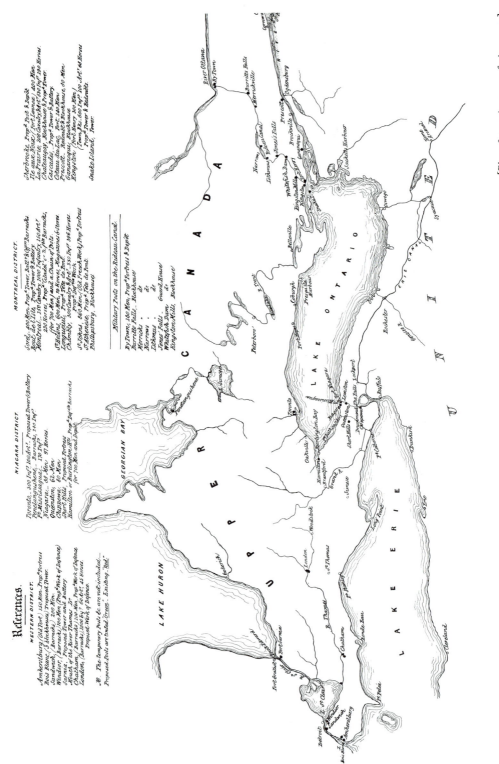

References.

WESTERN DISTRICT.

Amherstburg (Old Fort) 150 Men. Prop.ᵈ Fortress
Bois Blanc, (3 Blockhouses) Proposed Tower.
Sandwich, (Barracks) 200 Men.
Windsor, (Barracks) 100 Men. (Prop.ᵈ Work of Defence)
Sarnia, Proposed Tower and Battery
Reach of the River Thames
Chatham, (Barracks) 100 Men. Prop.ᵈ Work of Defence
London, (Barracks) 1000 by? 60 by?. as such
　　　　　　Proposed Work of Defence

N.B. The temporary Posts &c. are not included.—
Proposed Posts are tinted Green: Existing "Red."

NIAGARA DISTRICT.

Toronto, 200 Inf.ᵗ 100 Inf.ᵗ. Proposed Tower & Battery
Pendant que here...........Barracks, 240 Inf.ᵗ
Fᵗ. Mississauga,—　170 Inf.ᵗ
Niagara,— 80 Men. 97 Horses
Queenston, 62 Men.
Chippewa, 80 Men.
Fort Erie, Proposed Fortress
Short Hills, Proposed Fortress
Hamilton or Burlington Hᵗˢ. Prop.ᵈ Fortress
　　　　　　　for 700 Men and Rapids

MONTREAL DISTRICT.

Cornᵈ, 400 Men. Prop.ᵈ Tower, Batt.ʸ & Off.ʳˢ Barracks
Bout. de l'Ile,　Prop.ᵈ Tower & Battery
Montreal, 150 Cavalry 1000 Infantry. 150 Art.ʸ
230 Horses, Prop.ᵈ Citadel 'c' n.ᵈᵉ Barracks,
　　　　　for 700 Men, and a Chain of Posts.
St Helens, 500 Men, 10 Horses, Magazine 6 Stores
Longueuil, Prop.ᵈ Tête de Pont.
Chambly, 300 Cavalry & Inf.ᵗ 530 Inf.ᵗ 396 Horses
Chambly,　　　Prop.ᵈ De Work.
St Johns, 860 Men, Old French Work, Prop.ᵈ Fortress
St Athanase, Prop.ᵈ Tête de Pont.
Philipsburg, Blockhouse.

Sherbrooke, Prop.ᵈ Prit.ᵗ & Dep.ᵗ
Ile aux Noix, (Fort License) 400 Men.
La Prairie, 200 Cavalry Batt.ᵗ 530 Inf.ᵗ 190 Horses
Chateauguay, Blockhouse & Prop.ᵈ Tower
Cascades,　Prop.ᵈ Tower & Battery
Coteau du Lac,　Port, 150 Men.
Prescott, Blockhouse
Gananoque, Blockhouse
Kingston (Fort Henry 300 Men.)
　　　　　(Town Pts. 600 Inf.ᵗ 100 Art.ʸ 60 Horses
　　　　　Prop.ᵈ Tower & Redoubts
Snake Island, Tower.

Military Posts on the Rideau Canal.

By Town, 180 Men. Prop.ᵈ Fortress & Dep.ᵗ
Burritts Falls.. Blockhouse
Merricks .　.　do.
Narrows :.　.　do.
Isthmus .　.　do.
Jones Falls　Guard House
Whitefish Dam　do.
Kingston Mills, Blockhouse.

[Plate 3 continues on facing page]

[Plate 3, continued]

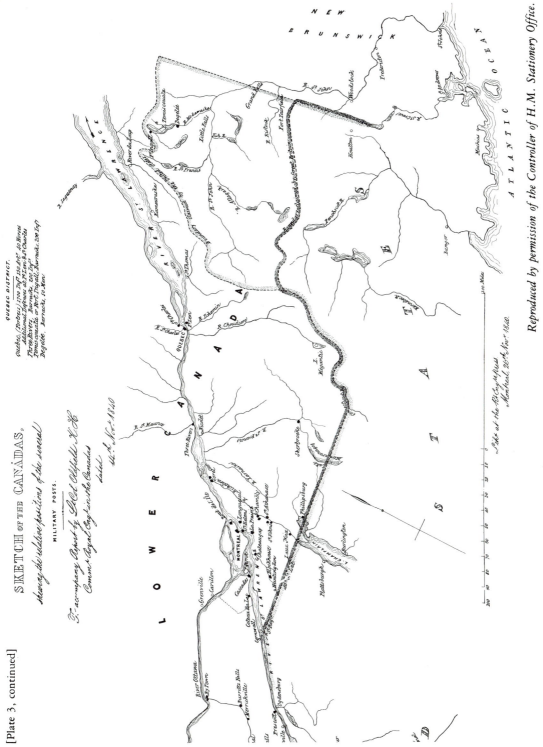

Plate 3 Sketch of Canada's defences by Lieutenant Colonel Oldfield of the Royal Engineers, November 14, 1840. Source: W.O.1/536.

Reproduced by permission of the Controller of H.M. Stationery Office.

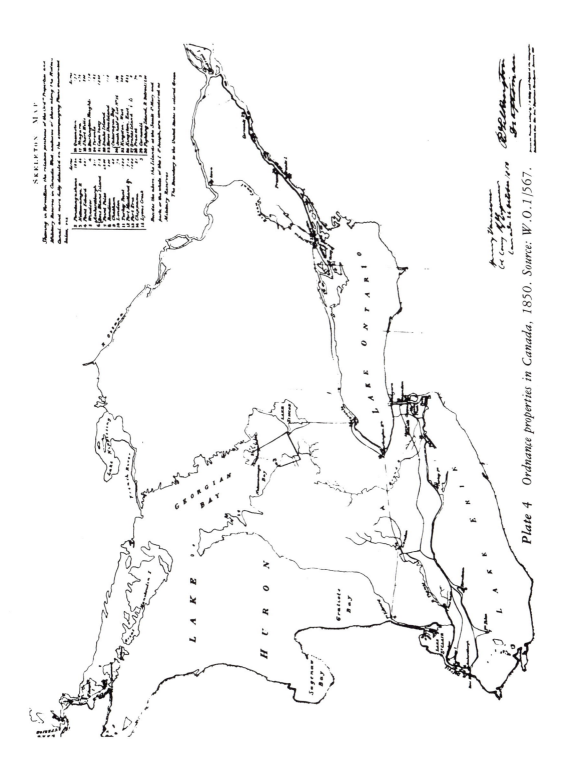

Plate 4   Ordnance properties in Canada, 1850. Source: W.O.1/567.

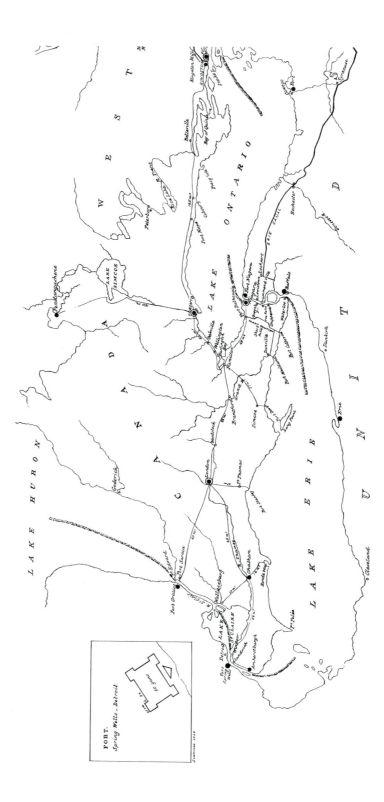

[Plate 5 continues on facing page]

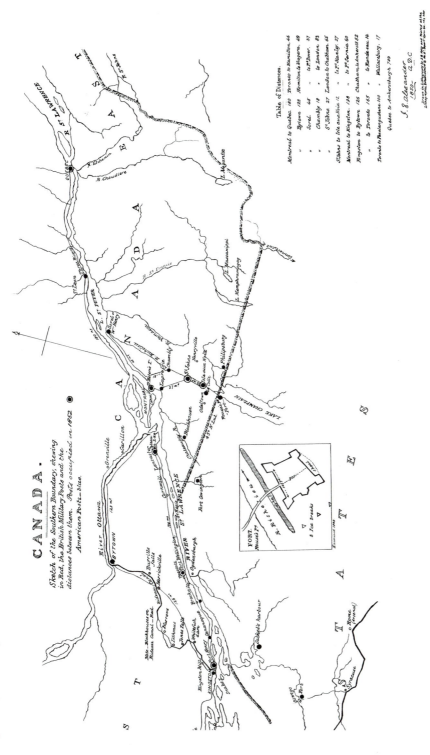

Reproduced by permission of the Controller of H.M. Stationery Office.

Plate 5   *Ordnance map of Canada's defences, 1852. Source: W.O.1/565.*

be the policy of Her Majesty's Government in regard to the gradual transfer, so far as may be practicable, of Ordnance and Military charges to the Provincial Government.[23]

There was, however, a new difficulty. Colonial secretary the Duke of Newcastle now sided with the Ordnance, claiming that only enough of the Rideau and Ottawa lands should be surrendered to meet any deficiencies in canal maintenance expenses. The Ordnance should retain the rest, placing income from rents into a fund for military projects.[24] The Treasury protested,[25] but when Newcastle refused to yield[26] fell in with his views.[27]

Unfortunately for the Ordnance it was too late to implement Newcastle's terms. Interdepartmental discord had given the initiative to the Canadians. In June 1854 J. S. Elliott, now Ordnance storekeeper in the province, reported that after discussions among himself, the governor general and leading members of the local legislature, the Canadians had decided to include the canals in the forthcoming general transfer of all Ordnance properties. While the general transfer was pending they agreed to pay canal maintenance costs, being confident they would get the lands in the near future.[28]

This general transfer scheme was the direct result of Earl Grey's campaign against the Ordnance; it also stemmed from Grey's indirect encroachment on the department's colonial properties. During the tedious canal negotiations Grey had been organizing military colonies on the department's lands in Canada West, and from 1851 his "enrolled pensioners" began to occupy the military reserves in major urban centres. The Ordnance resisted wherever it could, but found it impossible to prevent a continuing erosion of its Canadian holdings. The enrolled pensioner project was a component of Grey's total military reform effort.

In an attempt to ease colonial anxieties over the withdrawal of British troops, in 1846 Grey proposed the substitution of pensioned-off veterans for regulars in overseas garrisons. At the time annual pension payments in Great Britain stood at £1,191,350,[29] and the government was already assigning the maximum number of able-bodied pensioners to part-time police and garrison duties in order to get some value out of all this expenditure. Modelling his plan on the British system of enrolment, Grey set out to settle compact groups of pensioners of good health and character in or near colonial towns. These men, with their families, would be placed under the supervision of half-pay officers called staff officers of pensioners; once enrolled, they could support themselves by farming the acre or two granted to them, by taking jobs in their neighbourhood, and by means of their normal pension payments. In case of civil

23 Ibid., pp. 260-62, Trevelyan to Merivale, Nov. 3, 1853.
24 W.O.6/89, pp. 6-7, Merivale to Trevelyan, Jan. 2, 1854.
25 W.O.44/49, pp. 487-88, Treasury minute, Jan. 13, 1854.
26 W.O.1/568, p. 480, P. Smith to Trevelyan, Jan. 28, 1854.
27 Ibid., p. 486, Trevelyan to Merivale, Feb. 7, 1854.
28 W.O.44/49, pp. 502-03, Elliott to the R.O. Montreal, June 19, 1854.
29 C.O.885/1, Confidential memorandum on the state of the army, by Earl Grey, Oct. 17, 1846.

disturbance or military emergency they could be assembled very quickly, armed, uniformed and placed on active service. In this way Grey could reduce colonial defence expenditures—and the number of pensioners in Britain—while still being able to claim that he was providing British colonists with military protection. With the aid of the secretary-at-war, Fox Maule, and officers like Colonel J. D. G. Tulloch, he put this scheme into operation in New Zealand in 1847. When it succeeded, he turned to the other colonies.[30]

As usual, however, there were difficulties in Canada. Crown lands on which pensioners could be settled were no longer available. As Colonel Tulloch discovered, however, there were many unused acres of Ordnance reserves at Kingston, Bytown, Toronto, Hamilton, Chippewa, Fort Erie, London, Chatham, Amherstburg, and elsewhere. In 1851, without consulting the Ordnance, Grey sent 600 pensioners to occupy the reserves at Toronto, Penetanguishene, Amherstburg, Niagara, and London. He proposed to send more men when the first group was settled. The project was thus far a success, enabling substantial troop reductions in Canada West, giving useful employment to the pensioners and improving their standard of living. Unfortunately, in the midst of implementation an interdepartmental battle almost wrecked everything. The Ordnance, supported by the Treasury, refused to surrender the lands Grey and Fox Maule needed. Herman Merivale, the colonial undersecretary, later summed up the dispute as a most "vexatious case" and Newcastle called it an "obstinate" problem.[31]

Initially the Ordnance protested that it was impractical to hand over the reserves. For example, most of the Toronto lands had been leased to the municipal corporation. Inspector General John Fox Burgoyne thought that lands not already leased or rented should be sold rather than given up to pensioners who were unlikely to make good use of their small grants. The proceeds should go to defence works.[32] The Treasury disapproved of pensioner settlements on principle and argued with Burgoyne.[33] But the Colonial Office had assumed an obligation to provide the pensioners with land and, urged on by the War Office, issued repeated directives to the Ordnance to stop its obstruction. The directives were acknowledged but not carried out. In the meantime, various Canadian municipalities were making their own arrangements with the Ordnance and taking over the pensioner lands. Finally, in September 1853 all parties reached agreement after much agonized negotiation. The municipalities would get the lands and give the pensioners money instead.[34] In the end the pensioner lands were included in the general transfer arrangements for Ordnance properties and the provincial

30 For details see Raudzens, "A Successful Military Settlement," 389-403.

31 W.O.1/567, p. 154, Minute by Merivale, Feb. 12, 1853, and Minute by Newcastle, Feb. 12, 1853.

32 W.O.1/565, pp. 272-73, Memorandum by Burgoyne, Jan. 12, 1852.

33 Ibid., p. 252, Trevelyan to Merivale, Feb. 19, 1852, and p. 275, Trevelyan to Merivale, Feb. 21, 1852.

34 W.O.4/289, pp. 422-26, Hawes to Merivale, Sept. 17, 1853, and W.O.1/567, pp. 254-55, Trevelyan to Merivale, Oct. 26, 1853.

government assumed responsibility for paying compensation and employing the military settlers. The resistance of the Ordnance had only caused its officers further frustration.

Thus, by 1854 pressures from London and Canada were pushing the department out of the province. Only the final legal actions were necessary to divest it of its possessions. These took another two years. The transfer agreement was negotiated in London in the spring of 1854 by Francis Hincks, Canadian finance minister, Tulloch of the War Office, Newcastle and Trevelyan.[35] In September 1854 the new colonial secretary, Sir George Grey, drafted the detailed transfer plan. Ordnance property was divided into three classes: in the first class were the fortifications and barracks at Quebec, Montreal and Kingston, which would remain imperial stations manned by British regulars. Next were fortifications, barracks and canals transferred to the province on condition they were kept in military readiness. The third class included all remaining buildings and lands surrendered unconditionally.[36]

Everything now seemed settled but Ordnance officers still tried to preserve something for their department. Elliott protested that Grey's terms were unfair, pointing out that the Ordnance had been forced to purchase £205,947 worth of Canadian lands since the 1820s. It was one thing to give away properties inherited prior to 1826, but surely it was unjust to give up possessions purchased by British taxpayers for which the colonists had already been well paid.[37] Maintaining his opposition to the end, Elliott nevertheless was forced to partake in the final negotiations between representatives of the imperial government and Sir Alan MacNab and Colonel Etienne Tache of the Canadian executive.[38] After May 25, 1855, the Ordnance went out of existence; by the following December the transfer agreement was concluded.

By an act of the British Parliament, Ordnance properties were given up to the several colonies subject to confirmation by provincial acts. Only properties designated by Grey as imperial stations were retained.[39] On June 19, 1856, Canada's parliament confirmed the imperial statute[40] and on November 15 commanding Royal Engineer Colonel W. R. Ord reported that the transfer had gone into full effect ten days previously. Ord had turned over all deeds, plans and rent rolls to Major William F. Coffin, newly appointed provincial Ordnance land agent.[41] Taking only those properties in Grey's second class which were to be kept up for military use, Canada received 83,877

---

35 R.G.1, E 1, Vol. 78, pp. 31-33, Report of a committee of the Canadian executive council, Mar. 16, 1855.
36 W.O.1/568, pp. 140-45, George Grey to Elgin, Sept. 8, 1854.
37 W.O.44/49, pp. 569-71, Elliott to J. Wood, Mar. 20, 1855, and pp. 505-06, Return of Ordnance lands purchased with imperial funds, by Elliott, 1855.
38 Ibid., pp. 597-98, Edmund Head to Russell, June 4, 1855.
39 W.O.6/90, Mundy to Merivale, Dec. 8, 1855, and Imperial Statute 18 and 19, Vic. Cap: 117.
40 *Provincial Statues of Canada*, 19 Vic. Cap 45, 1856.
41 W.O. 44/720, Ord to Burgoyne, Nov. 15, 1856. The officers of the Royal Engineers and their commander, the inspector general of fortifications, came under the authority of the War Office in 1855.

acres valued at £345,000. The Ottawa-Rideau lands alone comprised 23,877 acres worth £80,000, producing an annual revenue of £6,255 in local currency.[42] The amount of property in Grey's third class was never clearly established but it was very much greater. Thus Canada gained a substantial inheritance from the Ordnance.

So ended the last unhappy decade of the department. Its officers pursued their duties with remarkable tenacity to the last, attempting to provide the colonists with adequate defences which would cost neither Britain nor Canada anything. This was a commendable aspiration, but it was not appreciated. Instead, the reformers in London dismantled the department because they considered it an instrument of government which had outlasted its usefulness, and the Canadians only saw it as an alien nuisance to be divested of attractive assets. Nevertheless it took ten years of concerted effort to eliminate this old and well-established department of state completely.

---

42 Ibid., Memorandum by Ord, Nov. 15, 1856.

Chapter X

The Ordnance Impact on Canada

Despite its many troubles the Ordnance probably contributed more to Canadian development than any other single military organization. Yet colonists seldom recognized the department itself as a benefactor; instead, they gave credit to individuals like Colonel By or to the armed forces generally. They heartily appreciated the money which poured into their communities from the Commissariat coffers, because the advantages derived from military spending on food, building materials, transportation facilities and all the other needs of the soldiers were most obvious. The garrisons also supplied free police services and their officers added lustre to the local elites. The colonists therefore were happy to have as many regulars as possible, and they protested vigourously against Earl Grey's withdrawal policy.[1] But the Ordnance did not attract such favourable attention.

The department's obscurity was in part a reflection of its peculiar situation within the imperial government. First, its work was often cloaked in official secrecy. Details of defence projects had to be kept from the all too proximate scrutiny of the potential American enemy. Thus the Canadian press did not appear to have noticed the work of the 1825 Smyth commission which was crucial to provincial security.[2] Second, while garrison soldiers were very conspicuous in small colonial communities, the engineers and sappers, relatively few in numbers, usually executed their tasks without much public notice. Third, the Ordnance was so isolated from the colonial authorities and Colonial Office representatives that the local inhabitants who wished to deal with it had to approach it indirectly. Finally, its tardy establishment in Canada in the mid-1820s left it without legal title to its properties until 1843. By the mid-1840s its most

---

1 See J. Mackay Hitsman, "Please Send Us a Garrison," *Ontario History*, L (1958), 189-91, and Philip, "The Economic and Social Effects of the British Garrisons," pp. 37-38.

2 For example, see 1825 issues of the *Montreal Gazette* and Kingston *Chronicle*. There is a brief notice in the September 14, 1825, issue of the *Canadian Courant* stating that Smyth and two other engineer officers have arrived in Halifax to examine British North American fortifications, but the editor makes no further comment.

vigourous period was over, and consequently its career in British North America was too brief to make much impression on the public.

As a result, while the Canadian press commented at length on the activities of Ordnance officers like John By and the geologist and author Captain Richard Bonnycastle,[3] the fact that they were members of a distinct department of state was overlooked. The *Bytown Gazette* in the late 1830s and early 1840s was exceptional; Bytown was too closely tied to the Ottawa-Rideau waterway to ignore the department. At the time of union between the upper and lower province, for example, the *Gazette's* editors and contributors pointed to the department's works to argue that Bytown should be the new capital of Canada. The Ordnance canals made the town more secure from American attack than Montreal, Kingston or Toronto, Colonel By's military reserves were thought to be ideal sites for government buildings, and Bytown was situated precisely on the boundary between French and English Canada. Similar arguments had been used by British soldiers since 1815 and by the Ordnance itself to promote the military waterway; Bytown had been planned as the major Canadian military depot. Yet in repeating these points the *Gazette* usually referred to Wellington, Dalhousie, John By or other authorities, not to the Ordnance.[4]

Two other matters aroused comment about the department in the *Gazette*—canal tolls and Ordnance land tenure. Heavy economic dependence on the waterway made increased tolls unpopular. But revised toll schedules were imposed by authority of the governors, not directly by the Ordnance, and while the department received its share of criticism, many of the complaints were directed at the representatives of the Colonial Office.[5] In the case of land, the *Gazette* from time to time attacked the department for obstructing economic progress by its refusal to surrender unused properties to civilians.[6] Here again, however, the attack only became serious after 1843, as a result of the Nicholas Sparks dispute. Up to 1843 the *Gazette* favoured the passage of the Ordnance vesting act as long as the just claims of Sparks and others were met. The newspaper welcomed the proviso restoring the Sparks property as a progressive step for Bytown.[7] Because the land in question was situated between what was called the upper and lower town, under departmental control it formed a barrier between the two developing sections of the community. When the Ordnance found that the proviso was defectively worded and reasserted ownership, the *Gazette* and its contributors castigated the department liberally with accusations of greed, breach of faith and conduct unbecoming a branch of the imperial government.[8]

---

3   Kingston *Chronicle*, Apr. 28, 1832.
4   *Bytown Gazette*, Apr. 18, 1838, Aug. 7, 1839, Aug. 27, 1840, Nov. 26, 1840, Jan. 14, 1841, Jan. 5, Feb. 16, Oct. 5 and Nov. 20, 1843.
5   *Bytown Gazette*, Nov. 24, 1842 and Feb. 2, 16, 1843. The merchants of other towns also complained about tolls on the Ordnance canals. See Toronto *Examiner*, Apr. 20, 1842.
6   *Bytown Gazette*, Nov. 19, 16, 1840.
7   Ibid., June 18, 1840 and Aug. 31, 1843.
8   Ibid., Feb. 15, 1844, Jan. 30, Mar. 6, 13, 20, 27, Apr. 17, May 1, 22, 1845.

For Bytown, therefore, the Ordnance was primarily an enemy. Yet simultaneously individual Ordnance officers were held in high regard. John By was a local hero,[9] and his successor, Major Donald Bolton, was much respected. The *Gazette* noted with approval Mrs. Bolton's social activities[10] and the Major's generous donation of a communion plate and a christening vase to Christ Church;[11] when a traveller publicly criticized Bolton's canal management, the *Gazette* printed a vigourous rebuttal.[12]

In addition to the neglect of the press, Canadian awareness of Ordnance contributions to provincial development was obscured by departmental opposition to local civil works projects.[13] From the 1820s Ordnance officers attempted to stop internal improvements which infringed on the department's lands, were too close to the American border, or involved north to south communications either toward or into the United States. For example, in November 1832 Richard Byham complained to the Colonial Office that Upper Canada had passed an act incorporating the Niagara Harbour and Dock Company, giving the company use of Ordnance reserves near Fort George. The department had received no notification of this in advance and Byham requested that the colonial secretary prohibit lieutenant governors from allowing such expropriations by the colonial legislature.[14] There was no immediate response. In May 1834 Byham renewed his complaints when the Upper Canadian legislature passed another bill, for a railroad from Lake Ontario to Lake Erie. On the insistence of Ordnance officers attempting to protect departmental land the lieutenant governor had reserved this bill. Byham asked for its disallowance at least until the respective officers found out exactly how far Ordnance interests would be affected.[15] Two years later the matter was still unresolved. The inspector general of fortifications stated it was impossible to determine the extent to which the railroad would encroach on Ordnance land because the company had not yet selected a route. As a precaution, however, he asked the Colonial Office to make sure that no civil work was constructed less than 1,000 yards from any fortification.[16] The colonial secretary directed the lieutenant governor to comply.[17] The department also managed to have a bill passed by the Upper Canadian legislature forbidding companies or individuals to appropriate military land without prior Ordnance consent.[18] But the resulting act was ineffective. In 1838 the department

---

9   In the editorial of the July 6, 1843, issue of the *Bytown Gazette* Colonel By is described as a financial genius among canal builders.

10 *Bytown Gazette*, Mar. 13, 1839.

11 Ibid., Sept. 26, 1839.

12 Ibid., June 18, 1840.

13 On the rare occasions when Canadian historians recognize the Ordnance they see it as an opponent of the communication projects designed to improve commerce with the United States. See Glazebrook, *History of Transportation in Canada*, I, 73-94.

14 C.O.42/412, pp. 76-77, Byham to Hay, Nov. 5, 1832.

15 C.O.42/424, p. 38, Byaham to Hay, Mar. 12, 1834.

16 W.O.47/1708, pp. 1862-65, Ordnance minute, Feb. 19, 1836.

17 W.O.44/42, p. 403, Stephen to Byham, Mar. 12, 1836.

18 Ibid., pp. 407-09, R. S. Jameson to John Joseph, July 7, 1836.

called for disallowance of the Kingston Marine Railway Company Act which again involved construction on Ordnance property. James Stephen at the Colonial Office argued that disallowance was not politically feasible and suggested that the department find some other way to safeguard its interests.[19] In the end the Ordnance had to settle for a compromise arrangement with the offending company.[20]

By the 1840s the department could no longer prevent Canadian entrepreneurs and corporations from expropriating its land. All it could do was attempt to preserve some of the military value of the surrendered properties. In 1848 when Colonel W. E. C. Holloway, C.R.E., found it impossible to reject the application of the Great Western Railway Company for permission to lay rails across the reserves at Burlington Heights, he insisted that no structures be erected which were likely to interfere with military works or operations.[21] The department was forced to make similar concessions to the city of Quebec in 1849,[22] to the Queenston Suspension Bridge Company in 1850[23] and to the Bytown to Prescott Railway and Toronto and Lake Huron Railway companies in 1851.[24]

The Ordnance opposed north-south communications projects more vigourously but with no better success. In 1832 it opposed the construction of the Chambly Canal on the Richelieu River. Colonel Gustavus Nicolls warned that this work would remove all natural obstacles between New York and Quebec, with the result that "New York may fairly be looked upon as a Base from which the Americans, under an energetick [sic] War Department, and enterprising Military commander, might, availing themselves of its resources, almost at once, operate first against Isle aux Noix, Montreal, etc., and subsequently Quebec."[25] The Ordnance wanted the Chambly Canal Act disallowed but acted too late to prevent royal assent.[26] It applied similar resistance to the St. Lawrence canals in the early 1840s and to projects aimed at bridging the Niagara River. In 1849 the respective officers tried to stop a provincial act incorporating the Queenston Suspension Bridge Company on the grounds that the bridge could be used by invading Americans. The act was reserved by the governor general but could not be disallowed.[27] Ordnance officers opposed the Fort Erie and Buffalo suspension bridge for similar reasons.[28] Lord Elgin temporarily reserved the company act but decided finally that its

19 W.O.44/33, pp. 140-42, Stephen to Byham, May 12, 1834.

20 W.O.47/1826, pp. 6949-54, Ordnance minute, May 31, 1839, and W.O.47/1828, pp. 7678-81, Ordnance minute, June 17, 1839.

21 W.O.44/46, p. 649, Holloway to military headquarters, Canada, Mar. 15, 1848.

22 R.G.1, E 4, Vol. 6, pt. 1, p. 207, Elgin to Grey, Feb. 25, 1850.

23 W.O.44/47, p. 340, H. Vavasour to Burgoyne, June 27, 1850.

24 W.O.44/48, p. 28, the R.O. Montreal to G. Butler, Aug. 27, 1851, and W.O.6/88, pp. 191-92, Merivale to Butler, Dec. 12, 1851.

25 W.O.44/42, pp. 42-47, Nicolls to Bryce, July 11, 1832.

26 Ibid., pp. 49-51, Memorandum by the master general, Sept. 22, 1832.

27 W.O.44/47, p. 330, The R.O. Montreal to Byham, Apr. 16, 1849, and pp. 346-47, Holloway to Burgoyne, June 14, 1849.

28 W.O.1/565, p. 173, The R.O. Montreal to Butler, Aug. 15, 1851.

value to the province outweighed its military disadvantages.[29] The Ordnance continued to resist until 1852, but the Colonial Office supported the Canadians and bridge construction proceeded.[30]

While perhaps justifiably opposing these kinds of projects, the department nevertheless gave the Canadians considerable direct aid where military interests were not involved. Canadians who wanted to build churches, for example, often received generous grants of Ordnance land, although as a rule contributions to actual church construction were not made. The only exception was in 1841 when the department gave £450 for the erection of a church at Sorel, on military land and partly for military use.[31] On the other hand, the Ordnance donated many church sites. In 1832 it gave a site at Grenville in Lower Canada,[32] in 1847 it granted land to the Church of England in Bytown,[33] in 1849 it did the same for St. Mark's Church at Niagara,[34] and in 1853 it presented the Anglican Bishop of Toronto with land for a new church at Bytown.[35] It made these grants on condition that the land was not needed for military purposes, that a sufficient proportion of the community—especially the poorer people—wanted a church, and that enough money had been raised by the prospective congregations to erect a church building.[36] In other respects the grants were entirely free.[37]

Similar aid was extended to charitable organizations, hospitals and schools. In 1843 the province decided to build a "Lunatic Asylum" at Toronto and the Ordnance agreed to supply the necessary twenty-five acres;[38] later the mayor and corporation thanked Sir Richard Bonnycastle of the engineers personally for this grant and also for other lands he had turned over to the city to be used as parks or sites for public improvements.[39] General R. D. Jackson and Colonel Holloway suggested that a condition should be attached to such grants; plans of all civilian buildings on Ordnance land should be examined by the respective officers to ensure that the structures could serve as fortifications in emergencies.[40] But nothing came of this proposal. In 1847 the Ordnance gave four lots in Bytown for a general hospital.[41] The following year it gave the district school trustees of Barrifield a part of the Kingston reserve for a school.[42] In 1853 a charitable society called "the Orphan's Home" received a piece of land at Victoria

29 W.O.1/564, p. 86, Elgin to Grey, Sept. 30, 1851.
30 W.O.1/565, pp. 180-83, Merivale to Pilkington, May 13, 1852.
31 W.O.44/47, p. 21, Memorandum by W. W. Hope, Feb. 5, 1850.
32 W.O.47/1589, pp. 9195-97, Ordnance minute, Oct. 22, 1832.
33 W.O.44/46, p. 614, The R.O. Montreal to Byham, Feb. 12, 1848.
34 Ibid., p. 552, Trevelyan to the Ordnance, May 21, 1849.
35 W.O.1/566, p. 123, P. Smith to General Rowan, Nov. 5, 1853.
36 W.O.44/49, p. 419, Minute by the inspector general of fortifications, Sept. 19, 1849.
37 Ibid., p. 411, Trevelyan to the Ordnance, Oct. 13, 1849.
38 W.O.44/44, pp. 111-12, Metcalfe to Lord Stanley, Apr. 6, 1843.
39 *The British Colonist*, Aug. 5, 1845.
40 W.O.44/44, pp. 126-28, Jackson to Stanley, July 26, 1843.
41 W.O.44/46, p. 604, Trevelyan to the Ordnance, June 3, 1847.
42 Ibid., pp. 541-44, Holloway to Burgoyne, Mar. 24, 1849.

Square, Toronto.[43] There were also instances when the department agreed to provide municipalities with park lands. In 1845 the bishop, mayor and "respectable inhabitants" of Toronto petitioned the Ordnance for permission to enclose a portion of the military reserves for use as a public park,[44] and the request was granted.[45]

Ordnance officers also helped to promote scientific enquiry in Canada. Their most notable contributions involved the establishment of meteorological observatories which by 1839 they were erecting throughout the empire.[46] In 1840 they set up a scientific station at Toronto, another at the Cape of Good Hope, and a third on St. Helena. [47] In 1847 Colonel Holloway drew up plans for a second observatory at Quebec,[48] and in 1850 it was built on the citadel. The Ordnance, Royal Navy and Royal Observatory at Greenwich provided the technical experts and instruments and the Canadians gave part of the money.[49]

Of the two Canadian establishments, the station at Toronto was most exclusively an Ordnance project and had the greatest impact on scientific study in the province. It was set up to investigate "Terrestrial Magnetism and Meteorology," and was staffed by a Royal Artillery detachment under Captain J. H. Lefroy.[50] A well-known Ordnance scientist, Colonel Edward Sabine, R.A., F.R.S., compiled the data gathered at Toronto and published it in three volumes;[51] the department viewed this project as a major success.

By 1852, however, sufficient experimentation was completed and the Ordnance decided to shut down the observatory. Captain Lefroy objected, urging his superiors to keep the institution in being on the grounds that it was of great value to science in general and of much benefit to Canada. It would materially assist the growth of the adjacent University of Toronto by stimulating the development of physics and mathematics departments. Lefroy therefore proposed that the observatory should be turned over to the university authorities intact.[52] A. N. Morin, the Canadian provincial secretary, supported this suggestion, stating that the province could undertake payment of all expenses if the Ordnance agreed to leave the building, instruments, and some of the Royal Artillery experts until such time as the university had the skilled staff to

43 W.O.1/567, p. 127, Colonial Office minute, by P. Smith, Nov. 12, 1853.
44 W.O.44/44, pp. 164-66, Holloway to Captain Talbot, Mar. 11, 1845.
45 W.O.44/49, p. 405, Stanley to Metcalfe, Apr. 18, 1854.
46 W.O.47/1837, pp. 11579-81, Ordnance minute, Sept. 18, 1839.
47 W.O.55/875, pp. 71-72, Trevelyan to the Ordnance, May 25, 1840.
48 W.O.6/88, pp. 35-37, Merivale to Butler, June 25, 1850.
49 Ibid., p. 26, Hawes to G. B. Airy, May 30, 1850.
50 Lefroy also conducted an important survey of the Lake Athabasca-Great Slave Lake region in 1843 and 1844 which improved scientific knowledge of magnetic disturbances and the position of the north magnetic pole. He founded the Canadian Institute, had a distinguished career in the artillery and colonial service, and was knighted in 1877. See *D.N.B.*
51 A full account of the origins and early operation of the Ordnance Observatories is given by Edward Sabine in his *Observations made at the Magnetical and Meteorological Observatory at Toronto in Canada* (London, 1845), pp. 9-19.
52 W.O.45/514, Memorandum by Lefroy, Nov. 19, 1852.

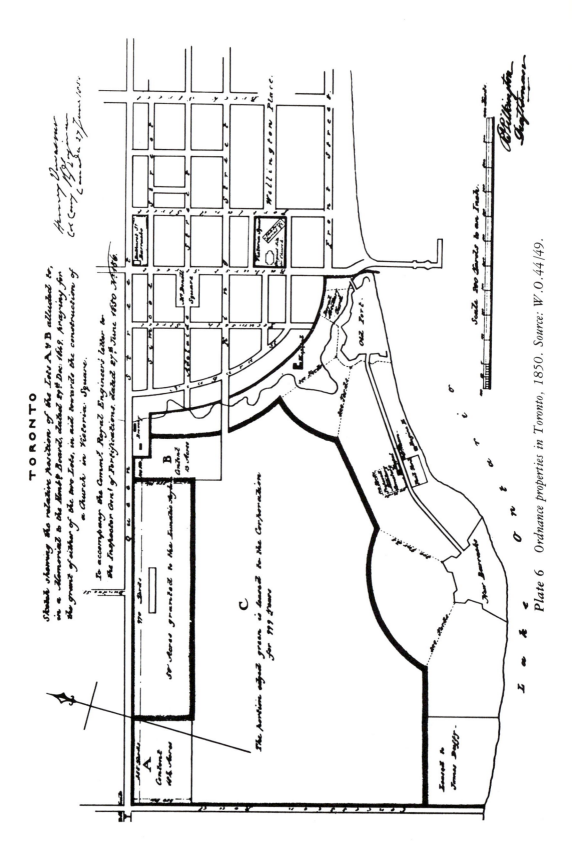

Plate 6   Ordnance properties in Toronto, 1850. Source: W.O.44/49.

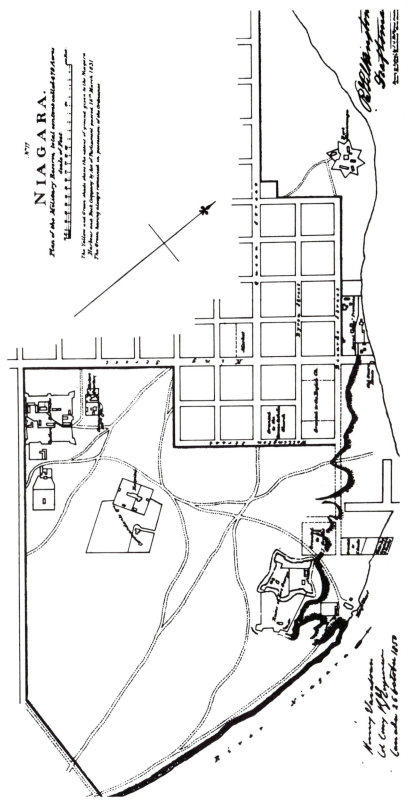

Plate 7 Ordnance properties in Niagara, 1850. Source: W.O.1/567.

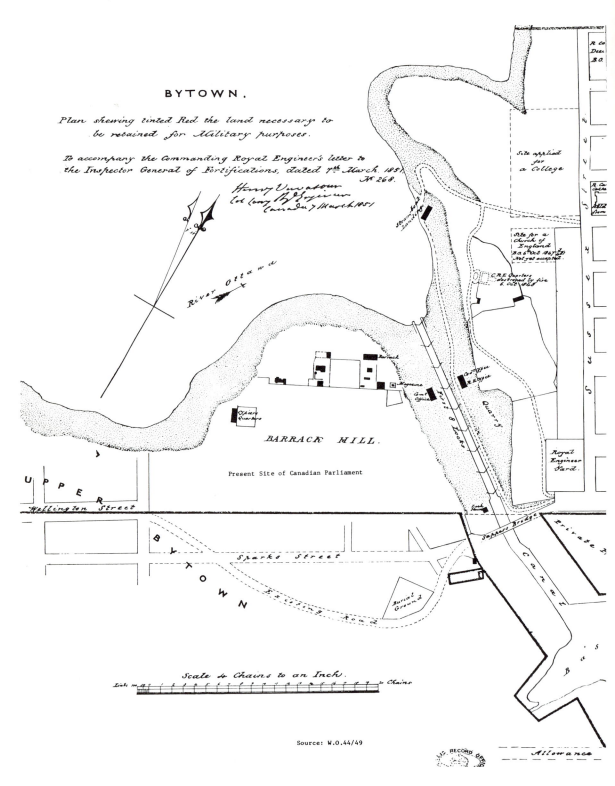

Plate 8   *Ordnance properties in Bytown, 1851*

replace them.[53] The Canadian executive council approved of this proposal[54] and other interests added their petitions, notably the Literary and Historical Society of Quebec, the Council of the Natural Historical Society of Montreal, the Canadian Institute and the Smithsonian Institution.[55] The latter argued that the Toronto observatory was essential to the proposed creation of a chain of similar research stations designed to span the whole hemisphere.[56]

After some hesitation the Treasury agreed to Lefroy's plan on condition that the province assumed responsibility for all costs.[57] On March 10, 1853, colonial secretary Newcastle announced that the imperial government would maintain the observatory as part of the University of Toronto.[58] The meteorological station which grew out of the Ordnance establishment remained in operation on the present site at 315 Bloor Street West adjoining the university almost until the present time.

Ordnance officers rendered equally valuable service in helping to explore the province. When the Duke of Wellington drafted his defence plan in 1819 he included among his recommendations the proposal for a waterway from the confluence of the Ottawa and Rideau rivers to Georgian Bay on Lake Huron via Lake Simcoe. Between 1819 and 1827 his officers surveyed the river systems along this proposed route to determine if they could be connected by canals and made navigable for military purposes. Lieutenants J. P. Catty, J. Walpole, W. B. Marlow and Henry Briscoe of the Royal Engineers examined the rivers, drew up maps and submitted reports of the topography and geology of a region previously known only to Indians and fur traders.[59] Lieutenant Catty, for example, travelled from Lake Simcoe to the Ottawa by way of the Talbot River and Madawaska, compiling the earliest accurate description of the part of the Canadian Shield now known as Haliburton.[60] The work of these engineers was followed up by Canadian geologists and land surveyors, helping to open the Ontario Shield country to lumbering, mining and agriculture.[61]

53   Ibid., Memorandum by A. V. Morin, Feb. 7, 1853.

54   W.O.1/566, pp. 22-23, Report of a committee of the Canadian executive council, Feb. 7, 1853.

55   Ibid., pp. 24-25, Elgin to Newcastle, Feb. 24, 1853.

56   Ibid., pp. 46-47, Smithsonian Institution to Elgin, 1853.

57   W.O.1/567, p. 165, Treasury minute, Mar. 4, 1853, and W.O.44/514, Trevelyan to the Ordnance, Mar. 7, 1853.

58   W.O.6/84, pp. 290-91, Newcastle to Elgin, Mar. 10, 1853.

59   James Watson Bain, "Surveys of a Water Route Between Lake Simcoe and the Ottawa River by the Royal Engineers, 1819-1827," *Ontario History*, L (1958), 15-28. See also C.O.42/368, p. 63, Maitland to Bathurst, Mar. 19, 1822, W.O.55/863, p. 277, Walpole to Melhuist, Oct. 22, 1825, pp. 349-59, Marlow to Wright, Sept. 22, 1826, pp. 372-77, Report by Briscoe, Oct. 16, 1826, W.O.44/18, pp. 118-19, Durnford to Wright, Oct. 14, 1826, W.O.55/863, pp. 378-81, Smyth to Mann, Dec. 2, 1826, W.O.46/29, pp. 31-32, Minute by Wellington, Jan. 10, 1827, and W.O.55/864, pp. 393-94, Smyth to Mann, Apr. 24, 1828.

60   Florence B. Murray, ed. and intro., *Muskoka and Haliburton, 1615-1875* (Toronto, 1963), Champlain Society Publication, p. xlv.

61   Don W. Thomson, *Men and Meridians: A History of Surveying and Mapping in Canada* (2 vols., Ottawa, 1966), I, 246-47.

Wellington's Georgian Bay waterway idea itself influenced the province. The citizens of Bytown and mercantile interests in Montreal and elsewhere became interested in it during the late 1830s, prevailing on the government of Upper Canada to organize further exploration. Captain F. H. Baddeley, R.E., worked with provincial surveyors in the search for a practicable route.[62] Such efforts ultimately resulted in the Georgian Bay ship canal project; following many decades of planning the Trent Valley Canal was constructed after 1907. This waterway connected Georgian Bay with Lake Ontario rather than the Ottawa River, but it followed a line first suggested by Wellington's engineers.[63]

Ordnance personnel were equally active in other types of survey work. Besides assisting in the determination of the Canadian-United States boundary,[64] they contributed to the geological exploration of Canadian resources. Work of this type by captains Bonnycastle and Baddeley was much appreciated by the inhabitants,[65] many of whom were eager to acquire as much reliable information about the physical structure of the province as possible. Some Canadians even regretted that imperial authorities did not do more such exploration.[66] Concerning land surveys, the Ordnance did not contribute as much in Canada as it did in newer colonies like South Australia and Van Diemen's Land.[67] By the time the department moved into British North America, Lower Canada was already settled and Upper Canada had enough civilian land surveyors to meet most of the needs of immigrants. But some Royal Engineers—notably F. H. Baddeley in 1835—did help with the more difficult surveying work, such as the marking of base lines from which boundaries of individual lots could be calculated.[68]

The outstanding Ordnance contribution, however, was the military waterway. From the inception of the Rideau project Canadians expected immense benefits from the canals. In the legislature of Upper Canada, in newspapers and in other ways they expressed their delight with Colonel By's efforts and the scale and durability of his work. They predicted that the Rideau Canal, aside from its military value, would do more to foster economic progress than anything yet attempted. They felt the deepest gratitude toward the mother country. In 1832 William Hamilton Merrit, whose Welland Canal had opened three years before, stated that the Ottawa-Rideau system would be such an effective counter to the competition of the New York Erie Canal that the Montreal

---

62 *Bytown Gazette*, June 21, Nov. 1, 1837 and May 27, Oct. 14, 1841. Richard Henry Bonnycastle also assisted with this work. For additional details see *Archives of Ontario*, Macaulay Papers, Baddeley to John Macaulay, May 30, June 7, July 27, Sept. 4, Sept. 15, Sept. 20, Sept. 26, Oct. 2, 1837, and House of Assembly, Upper Canada, "Report of the Commissioners on the Survey of the Ottawa River, Etc.," by John S. Cartwright, Mar. 27, 1839.

63 Glazebrook, *History of Transportation in Canada*, II, 225-26.

64 Connolly, *History of the Royal Sappers and Miners*, pp. 347, 416, and Thomson, *Men and Meridians*, I, 258-63.

65 Kingston *Chronicle*, Apr. 28, 1832, and *Montreal Gazette*, Apr. 2, 1837.

66 *Bytown Gazette*, Nov. 2, 1843.

67 Connolly, p. 342, and W.O.6/101, pp. 92-96, Merivale to Butler, July 11, 1851. See also entry for E. C. Frome in the *Australian Dictionary of Biography*.

68 Thomson, *Men and Meridians*, I, 247.

merchant community would gain ascendency over the trade of the American Middle West. Thus Canadians would dominate the interior economy of the continent.[69]

Until the St. Lawrence canals went into operation in 1848 many of these expectations were in fact fulfilled. Notwithstanding the bottleneck at the Grenville Canal, which was never completed on By's enlarged scale, and despite the Ottawa Steamboat Company's monopoly over St. Ann's canal at the mouth of the Ottawa until 1841, the military water route served as the main Canadian communication from the sea to the Great Lakes.[70] To ship goods up the St. Lawrence required twelve days and much labour; only small craft like Durham boats or bateaux could navigate the swift currents. By contrast 200-ton steamers could pull strings of large barges up the Ottawa-Rideau system in five days. Downstream shipment of bulk exports remained more economical on the St. Lawrence despite cheaper rates on the military canals. Grain and flour carriers went from Kingston to Montreal in less than three days via the river, and more than half the barges towed up the canals returned down the St. Lawrence. Even here, however, the canals provided the barges for the down traffic, and were almost indispensable to the upward traffic.[71] For example, in 1835, before the military route reached full capacity, superintending officer of the Rideau Canal, Major Donald Bolton of the engineers, estimated that more than three-quarters of all imports into Upper Canada—over 8,000 tons a year—passed through the military waterway.[72] That proportion increased as the triangular Montreal-Bytown-Kingston traffic expanded. In 1837, a bad year for trade, five steamers and twenty barges brought up 4,500 tons of imports; in 1839 the tonnage went up to 12,000 and in 1840 to 15,000. New Yorkers began to complain. They blamed the Rideau Canal, supplying barges for the down traffic, for diverting western products exceeding a million bushels of grain and flour from the United States. Also in 1840, 2,400 tons of bulk Liverpool salt came up the canal, depriving Oswego salt merchants of £20,000 worth of customary exports to the Canadians.[73] In the peak year 1844, when the military waterway was meeting all the needs of Canadian commerce, 35,000 tons of imports passed through the canals and 609,000 bushels of flour, 360,000 bushels of wheat and 21,000 barrels of pork arrived at Montreal.[74]

The timber trade derived similar benefits. Even during the construction stage the canals facilitated the floating of timber rafts; afterwards it became possible for the first time to exploit the vast forest resources of the Rideau basin. From the mid-1830s—with an interruption during the rebellion crises—the sawn timber trade catering for American markets expanded rapidly along both the Ottawa and Rideau arms of the waterway.[75]

---

69 Sneyd, "The Role of the Rideau Waterway," pp. 29-31 and 42-43.
70 Ibid., pp. 61, 69, 78, 81.
71 Ibid., pp. 61-63.
72 W.O.44/42, pp. 61-70, Bolton to the R.O. Quebec, Feb. 20, 1836.
73 Sneyd, "The Role of the Rideau Waterway," pp. 66-68.
74 Ibid., p. 121.
75 Ibid., pp. 47, 53.

As well, the immigrant traffic generated by the military canals helped to speed the growth of population. While travel along the St. Lawrence was cheaper, it was far more comfortable to use the Ottawa-Rideau system, and from 1834 the number of immigrants increased annually. By 1840, 12,000 people passed through the locks at Bytown during the navigable season; of these 1,400 settled in the Bytown area and 6,000 came under assisted passage. Therefore, until 1846 the military waterway was the principal funnel of new population into Upper Canada.[76]

Besides this the canals accelerated the general economic growth of present-day eastern Ontario. Local enterprise was stimulated by the need to organize shipping companies. In 1841 six such "forwarding" companies were busy on the waterway, among them McPherson and Crane Company (with eleven steamers and forty-five barges), Hooker, Henderson and Company, Sanderson and Murray Company, H. and S. Jones Company (ten steamers and ninety barges), and two others.[77] Flour, saw, carding, grist and oatmeal mills sprang up along the canals, benefitting from Ordnance works which improved the control of water levels and flow rates. New markets and settlement areas opened and land values rose.[78] Existing towns expanded and new ones appeared; Richmond, Kemptville, Burritt's Rapids, Smiths Falls, Perth, Newboro, Jones Falls, Seeleys Bay and other communities all increased in size and prosperity. The population of Kingston rose from 3,500 in 1830 to 12,000 in 1848.[79] In Perth the citizens were so drawn by the commercial opportunities presented by the Rideau Canal they determined to build their own canal to gain access to this waterway. In 1834 eleven miles of the Tay River—connecting Perth to the Rideau system—were navigable for steamers; the effort was a striking example of the secondary benefits caused by military expenditure.[80]

Perhaps Bytown experienced the biggest impact. According to one Canadian newspaper the community was founded "in honor of Colonel By."[81] When the founder had arrived in 1826 there had been only a handful of settlers on the site. The population grew in direct response to the needs of canal construction, to demands for labour, food, building materials and other staple commodities. The Colonel himself took a paternal interest in town planning, laying out streets, providing basic amenities and setting aside ground for public buildings. In 1832 population was in excess of 3,000; in 1842 the citizens succeeded in making their town administrative capital of the District of Dalhousie.[82] In 1855 Bytown became the city of Ottawa and in 1861, with a population of 14,669, it was the fourth city in Canada, exceeded in size only by Montreal, Quebec and Toronto.[83]

---

76 Ibid., pp. 55-56 and 58.
77 Ibid., p. 96.
78 Ibid., p. 164.
79 Ibid., pp. 205-06.
80 Ibid., pp. 181-83.
81 Kingston *Chronicle*, May 9, 1827.
82 Sneyd, "The Role of the Rideau Waterway," pp. 170-74.
83 W. L. Morton, *The Critical Years: The Union of British North America, 1857-1873* (Toronto, 1964), p. 3.

The Ordnance also helped to make it Canada's capital. When the colonists came into direct contact with the Ordnance they were usually hostile. Bytown's residents opposed the department more persistently than most Canadians. Ordnance tenants and men like Nicholas Sparks, William Draper and Stewart Derbishire used political weapons against the department which became part of a larger campaign for provincial self-government. Conversely, during the late 1830s and early 1840s much of the Bytown desire to make their town Canada's capital was due to varied Ordnance influences on the growth of the community. Even military and strategic Ordnance arguments were adopted for this cause. In 1857, when provincial politicians found they could not agree which of the three cities, Montreal, Quebec, or Toronto, should be the capital, they petitioned Queen Victoria to make the choice. In doing this the monarch was influenced not only by the need to select a place least likely to offend either French or British Canadians but also by the old military arguments about the Ottawa-Rideau canals. Possibly the decisive document which swayed the Queen was penned by the Canadian governor-general, Sir Edmund Walker Head, who put the following case among his various points:

> In a military point of view (I speak of course with submission to higher authorities), Ottawa is advantageously situated. Its distance from the frontier is such as to protect it from any marauding party, or even from a regular attack, unless Montreal and Kingston, which flank the approach to it, were previously occupied by the enemy. Stores and troops could be sent to Ottawa either from Quebec or Kingston, without exposure on the St. Lawrence to the American frontier.
>
> A secondary consideration, but one of some importance as affecting the popularity of the choice, is the fact that the Rideau Canal, now handed over to the Provincial Government, would probably increase its traffic and become more productive by the transfer of the seat of Government to Ottawa.[84]

Today the nation's parliament and government offices stand on what was once Ordnance land, which officers of the department kept clear of settlement long enough to make it available for its present uses.

Queen Victoria's choice had much effect on Canadian politics. In the short run it removed a heated dispute from a Canadian legislature so beset with antagonisms that its ability to carry on the business of government was in serious doubt. To that extent the choice allowed concentration on more important issues and may have helped Canadians along the road to confederation. In the long run Ottawa's location has assisted politicians in their perennial struggle to find acceptable compromises between their French-speaking and English-speaking electorates.

The peacetime benefits of these canals, then, were more pronounced than their military effects. Of course there was never the test of war. Yet it has been postulated that all the leading soldiers from Wellington down who held the waterway as the key to

---

84 James A. Gibson, "Sir Edmund Head's Memorandum on the Choice of Ottawa as the Seat of Government of Canada," *C.H.R.,* XVI (1935), 411-17.

Canada's survival were possibly more justified than they imagined. The Ottawa-Rideau system may well have acted as a deterrent to those Americans persistently devising plans of conquest and annexation.[85]

Despite all this, Canadians, and especially the residents of Ottawa, have not given credit to the Ordnance; the department has been forgotten, They revere, instead, one of its most troublesome officers, John By. This, of course, is understandable. He built his canal on a scale more grandiose than London wanted and he certainly promoted the growth of the community which bore his name until 1855. It is also easier to honour an individual rather than a long defunct organization. Therefore Ottawa, the Engineering Institute of Canada, the federal Historic Sites and Monuments Board, the Federal District Commission, the Royal Engineers, the Royal Canadian Engineers and sundry writers celebrate By's achievements in statuary, on plaques, in place names and in prose.[86] His achievements are inflated to the proportions of legend while the imperial body of which he was but a temporary instrument is unrecognized. Yet the myth of John By, the hero of eastern Ontario, is in itself an indirect Ordnance effect on Canadian life.

One more major Ordnance influence deserves notice. Not counting in excess of one million pounds sterling spent on the Ottawa-Rideau waterway, the department disbursed £2,769,005 in Canada between 1825 and 1843.[87] As a sample, during the financial year 1843-44 Ordnance expenditure amounted to £85,452. In the same period Britain's total net expenditures in Canada were £557,950, of which £541,019 covered military costs, £708 was spent for the navy and £16,222 for civil expenses; thus Ordnance spending amounted to 15.3 per cent of the total capital input of the imperial government.[88] In other years the percentage was higher. In 1830 the department alone spent £399,532 and Colonel By spent another £200,000, a sum in excess of half a million.[89] While the precise proportion of British money spent in Canada by the Ordnance between 1825 and 1855 is hard to esimate due to incomplete records,[90] it was certainly large and of great value to a pioneer region almost entirely dependent on external capital for growth. The money provided work and stimulated agriculture and manufacturing. Specifically, Ordnance spending gave jobs to poorer immigrants who were often dependent on canal and fortification projects for wages which could give them sufficient savings to purchase farms or go into business. Much of the department's money also went to civilian contractors, accelerating the growth of Canadian building enterprises. Most important, this money, unlike private capital, was interest free, and

85   Sneyd, "The Role of the Rideau Waterway," pp. 133-38.

86   Legget, *Rideau Waterway*, pp. 218-19. Also luncheon talks with officers of the Royal Engineers at the Royal Engineers Mess at Chatham, England, during the summer of 1966. In particular, discussions with Colonel E. E. N. Sandeman, Corps Librarian.

87   C.O.42/527, pp. 21-22, British military expenditures in Canada, 1825-1834, and P.P. 1844(304) XXXIV, 23, British military expenditure in Canada, 1835-1843.

88   P.P. 1849 (224) XXXIV, "Colonial Military Expenditures."

89   C.O.42/257, pp. 21-22, British military expenditures in Canada, 1825-1834.

90   Stacey, *Canada and the British Army*, pp. 40-43. In his calculations of British military spending in the colonies, Colonel Stacey notes that complete financial records are not available.

Ordnance money as distinct from other forms of government spending produced a wide range of long term benefits additional to the effects of capital influx alone.

Ordnance funds were often used to purchase land and build defence works which were given to the province free of charge after 1855. Canadians got both the money and the proceeds. In a few cases individual colonists made profits by selling land to the Ordnance and then repossessed the land itself after its value had increased. The property Nicholas Sparks regained in the 1840s was then worth so much it became the business centre of present-day Ottawa.[91] The total quantity of land the Ordnance surrendered was never precisely calculated but it was at least several hundred thousand acres, most of it very valuable because it was in urban areas. The inhabitants of the Rideau region alone received 23,877 acres worth £80,000 and yielding £6,255 local currency in annual rents.[92] Other places gained similar if smaller legacies and derived both social and economic benefits from them. In cities such as Quebec, Ottawa, Kingston, Toronto and London the Ordnance kept grounds clear of civilian occupants until municipal authorities were sufficiently mature to want luxuries like public pleasure grounds. Canadians therefore owe the Ordnance a debt of gratitude for many city parks such as Kingston's John A. Macdonald Park.[93] Finally, even the Ordnance defence works, no longer of any military use, have become an asset. Old Fort Henry, the Citadel at Quebec, and other fortifications today bring in substantial revenues from the tourist trade. Again, the Rideau Canal is the major attraction for holiday makers from the United States as well as Canada.

*   *   *   *   *

The military contributions to Canada's growth after 1815, then, were indeed considerable, and the Ordnance was responsible for most of them. Scholars have generalized about the beneficent effects of the British military presence, about expenditures, police services, architectural trends, the social influence of soldier settlements, the presence of officers and gentlemen in pioneer communities, and even styles of Canadian art. Yet how can these broad influences be quantified or documented? Is it worth trying to track down every instance where the architectural design of a colonial structure was shaped by an officer trained in Britain's military academies? Is it necessary to unravel the confused complexities of Britain's military accounts to find exactly where every shilling was spent? How can the subtle cultural or psychological pressures exerted by the officer class on the upper ranks of colonial society be weighed? These questions may still be worth pursuing; there is certainly room for more detailed study of the

---

91 Legget, *Rideau Waterway*, p. 205.

92 W.O.47/720, Memorandum by Colonel W. R. Ord, Nov. 15, 1856.

93 W.O.44/49, p. 621, Ordnance plan of Kingston, Jan. 7, 1853. The plan shows that the present area of Macdonald Park, the adjoining lake front and a large part of the site of Queen's University were all Ordnance property. See also W.O.44/44, pp. 164-66, Holloway to Captain Talbot, Mar. 11, 1854, and R.G.1, E4, Vol. 6, pt. 1, p. 207, Elgin to Grey, Feb. 25, 1850, for examples of parks in Toronto and Quebec situated on Ordnance land.

relationship of soldiers and civilians at the local level in pre-confederation English Canada. In the meantime, at least the Ordnance story indicates the overall magnitude of the military influence because this forgotten department contributed most of the benefits to colonial life which from time to time have attracted some attention.

Not only does the study of the department clarify the total size of the effect but it also shows how and why the effect occurred. Some soldiers were conscious benefactors. Certainly Colonel By was one of these, although it can be argued he did more harm than good by antagonizing those who controlled the flow of military funds to the colonies. For the most part the activities which promoted progress in Canada were unintentional or incidental to military purposes. There were more officers who tried to use civilian reasons for military projects than there were men like John By; such projects, too, often were not in the interests of civilian progress. Essentially the Ordnance—and other military organizations—sought its own narrow goals in the face of Canadian opposition and criticism from the British parliament. The Canadian defence problem was a troublesome responsibility to all imperial authorities, especially to a department under constant assault from reforming elements in the House of Commons, elements who saw the Ordnance as an intolerably anomaly and viewed colonies as burdens to be shed, ultimately, by extending the grant of self-government. In these unhappy circumstances soldiers had to extract funds for essential defences by methods bordering on fraud. It is little wonder that even the contributions of so potentially effective an instrument of civilian development as the Ordnance had only a marginal influence on the economic and social growth of Canada. The civilizing impact of Britain's legions was but a pale reflection of the works of Rome. But the impression left by a maligned and constrained Ordnance was still unique and impressive. The provinces of Quebec and Ontario owe much more than their inhabitants know to this imperial department. They are beneficiaries of one of the outstanding military organizations of the modern era which did more for the pursuits of peace than war.

Bibliography

This study is based primarily on the official records at the Public Record Office in London, at the Public Archives of Canada in Ottawa and at the Scottish Record Office in Edinburgh. The most important P.R.O. record series is War Office 44 (W.O. 44) but many other series had to be consulted—especially W.O. 1, 4, 6, 46, 47 and 44, C.O. 42, the P.A.C. Record Groups 1, 7, and 8, and the Scottish Record Office Government Documents (G.D.) series 45. For the Rideau Canal there are two useful sets of British Parliamentary Papers, 1830-31 (135) IX, 15, "Canada Canal Communication" and 1831-32 (in 570) V, Report of a select committee of the House of Commons on Canal Communication in Canada, June 29, 1832. These have been used for the sake of convenience but they consist almost entirely of documents at the P.R.O. which have been examined in their original form.

Because the Ordnance was dissolved in 1855 there is no specific group of papers dealing with the department. It appears from statements on title pages of the volumes of documents that War Office and P.R.O. clerks sorted and compiled the contents after the department's disestablishment; they appear to have followed no consistent pattern of classification. Thus the W.O. 44 and W.O. 1 series contain material on every sort of military activity, sometimes arranged chronologically, sometimes topically and sometimes geographically. Only some of the W.O. series like W.O. 4 (Secretary at War, Out-Letters) and W.O. 47 (Ordnance Minutes) are logically organized. Nor are there complete indexes to the W.O. papers. For any given event all the series as well as documents in the C.O. series and collections in places like the P.A.C. must be searched. Some documents are also lost. Covering letters referring to enclosures but without them and enclosures without covering letters are common. While it is known that some of these missing items were removed by masters general for private files, they are difficult to trace. Private papers of secretaries and clerks often in the best position to know about departmental activities have largely disappeared; neither Seth Thomas nor Richard Byham, for example, appear to have left personal records. There is little biographical information about most of the men who served the Ordnance in Canada.

While secondary sources are useful for background and context they contain little on the Ordnance as such. After much searching in the British Museum, the National Library of Scotland, the Royal Engineers Corps Library at Chatham, the Library of the Royal United Services Institution at London, the War Office Library, the Institute of Historical Research at the University of London, the library of the Royal Military College at Kingston, Ontario, and numerous university libraries, the results were meagre. Only a thorough investigation of all available British military papers between 1815 and 1855 has made this work possible.

Primary Sources

1.   Public Record Office, London.

W.O. 44   Ordnance In-Letters. These documents also deal with the military activities of other departments. Over 700 volumes in this series were examined. The most useful ones, particularly in regard to Ordnance canals, were W.O. 44/15 to W.O. 44/49 inclusive.

W.O. 1   War Office In-Letters. The most useful volumes were those between W.O. 1/500 and W.O. 1/600.

C.O. 42   Colonial Office papers relating to British North America.

W.O. 55   Ordnance Miscellanea.

W.O. 4   Secretary at War. Out-Letters.

W.O. 6   Secretary of State. Out-Letters.

W.O. 43   Commander in Chief. Out-Letters.

W.O. 46   Ordnance Out-Letters.

W.O. 47   Ordnance Minute Books.

W.O. 33   Inspector General of Fortifications papers.

W.O. 28   Head Quarters Record.

W.O. 25   Officers' Services.

W.O. 26   Miscellany Books.

C.O. 885/1   Cabinet papers, printed for the Colonial Office.

T. 28   Treasury Out-Letters, Naval and Military.

2.   Public Archives of Canada.

R.G. 8   British military and naval records relating to British North America. These papers are copies or duplicates of the W.O. series, particularly W.O. 44 and W.O. 1. Occasionally they contain enclosures missing from the P.R.O. collection.

R.G. 1   Executive Council papers, Province of Canada.

R.G. 7   Governor General's Office papers.

3.   Scottish Record Office, Edinburgh.

G.D. 45   Dalhousie Muniments. This is Lord Dalhousie's collection of official records.

4.   Scottish National Library, Edinburgh.

Sir George Murray MSS.

5.   Apsley House, London.

Wellington MSS.

6.   Royal Engineers Corps Library, Chatham.

T. W. J. Connolly Papers. Notitia Historica of the Corps of Royal Engineers. This document contains brief biographical sketches of Royal Engineer officers.

7.   Archives of Ontario, Toronto.

Nicholas Hugh Baird Papers.
Dr. H. T. Douglas Papers.
John Macaulay Papers.
J. E. R. Munro, "Sir Richard Henry Bonnycastle, Lieutenant-Colonel Royal Engineers."

Printed Documentary Sources

British Parliamentary Papers, 1826-27 (161) (380) XV, 285, 283, "Water Communication in Canada."
————— , 1830-31 (135) IX, 15, "Canada Canal Communication."
————— ,1831-32 (in 570) V, Report of Select Committee of the House of Commons on Canal Communication in Canada, June 29, 1832.
————— , 1833 (543) XXVI, 229, Report of two Commons Select Committees on the Phillpotts Case, July 8, 1833.
————— , 1844 (304) XXXIV, 23, "Return of the Charges incurred on account of the Canadas, in respect of the Army, Navy, Ordnance, and Commissariat, in each of the Years 1835 to 1843 inclusive. . . ."
————— , 1849 (224) XXXIV, "Colonial Military Expenditures."
Brymner, Douglas, ed., *Report on Canadian Archives*, 1890. Ottawa, 1890: Note D. Internal Communications in Canada, pp. 67-96.
————— . *Report on Canadian Archives*, 1897. Ottawa, 1897: Note C. Internal Communications in Canada, pp. 57-85.
Doughty, Sir Arthur C., ed. *The Elgin-Gray Papers, 1846-1852,* 4 vols. Ottawa, 1937.
House of Commons Journal, 1831-1832.
Provincial Statutes of Upper Canada. 1815-1841.
Provincial Statutes of Lower Canada. 1815-1841.
Provincial Statutes of Canada. 1841-1856.
Royal Engineers, The Corps of. *Papers and Subjects Connected with the Duties of the Corps of Royal Engineers.* London, 1837.
Royal Engineer Institute. *Corps Archaeologia.* Chatham, England, 1895.
Wellesley, Arthur, ed. *Despatches, Correspondence, and Memoranda of Field Marshal Arthur, Duke of Wellington, K.G.* 4 vols. London, 1867.

Newspapers

British Colonist (Toronto). 1838-1855.
Brockville Recorder. 1824-1849.
Bytown Gazette. 1836-1845.
Canadian Courant (Montreal). 1815-1829.
Chronicle (Kingston). 1819-1832.
Chronicle and Gazette (Kingston). 1833-1845.
Examiner (Toronto). 1840-1855.
Globe (Toronto). 1846-1855.
Hamilton Spectator. 1847-1855.
Montreal Gazette. 1815-1855.

Ottawa Citizen. 1851-1855.

Packet (Bytown). 1846-1850.

Patriot (York). 1829-1839.

Toronto Patriot. 1840-1853.

U.E. Loyalist (York). 1827-1828.

Upper Canada Gazette (York). 1815-1829.

Secondary Material. Books and Articles

Aitken, Hugh G. J. *The Welland Canal Company. A Study in Canadian Enterprise.* Cambridge, Mass., 1954.

Allen, James de Vere. "Two Imperialists: A Study of Sir Frank Swettenham and Sir Hugh Clifford." *Journal of the Malayan Branch, Royal Asiatic Society* (Singapore), XXXVII (1964), 41-73.

Bain, James Watson. "Surveys of a Water Route between Lake Simcoe and the Ottawa River by the Royal Engineers, 1819-1827." *Ontario History,* L (1958), 15-28.

Baker, Edward John. *Observations on the Rideau Canal.* Kingston, U.C., 1834.

Barraclough, G. *The Origins of Modern Germany.* Oxford, 1966.

Beaglehole, J. C. "The Colonial Office, 1782-1854," *Historical Studies. Australia and New Zealand,* I (1940), 170-89.

Beers, Henry Putney. *The Western Military Frontier, 1815-1846.* Philadelphia, 1935.

Billings, C. E. "The Rideau Canal," *Women's Canadian Historical Society of Ottawa, Transactions,* II (1909), 46-55.

Blanco, Richard L. "Reform and Wellington's Post Waterloo Army, 1815-1854." *Military Affairs,* XXXIX (1965), 123-31.

Blue, C. S. "John By: Founder of a Capital," *Canadian Magazine,* XXXVII (1912), 573-79.

Bonnycastle, Sir Richard Henry. *The Canadas in 1841.* 2 vols. London, 1842.

Bourne, Kenneth. *Britain and the Balance of Power in North America, 1815-1908.* London, 1967.

Brault, Lucien. *Ottawa Old and New.* Ottawa, 1946.

Burroughs, Peter. "The Search for Economy: Imperial Administration of Nova Scotia in the 1830's." *Canadian Historical Review,* XLIX (1968), 24-43.

Careless, J. M. S. *The Union of the Canadas. The Growth of Canadian Institutions, 1841-1857.* Toronto, 1967.

Connolly, T. W. J. *History of the Royal Sappers and Miners.* 2 vols. London, 1855.

Corey, Albert B. "Canadian Border Defence Problems After 1814 to their Culmination in the Forties." Canadian Historical Association, *Report* (1938), pp. 111-20.

Cornell, P. G. "William Fitzwilliam Owen, Naval Surveyor." *Collections of the Nova Scotia Historical Society,* XXXII (1959), 161-82.

Cowan, C. D. *Nineteenth-Century Malaya; The Origins of British Control.* London, 1961.

Cowan, Helen I. *British Emigration to North America, 1783-1837.* Toronto, 1928.

Craig, Gerald M. *Upper Canada; The Formative Years, 1784-1841.* Toronto, 1963.

Creighton, Donald. *The Commercial Empire of the St. Lawrence, 1760-1850.* Toronto, rev. ed. 1956.

Curry, F. C. "The Rideau Canal System." *Inland Seas,* XXI (Fall, 1965), 210-23.

Davies, Godfrey. *Wellington and His Army*. Oxford, 1954.

Drummond, A. T. "Some Notes on the Rideau Canal, the Sources of its Water Supply, and its Early History." *Canadian Record of Science*, V (1893), 459-71.

Dunabin, Thomas. "Side-lights on the Ottawa-Rideau Waterway to Upper Canada." *Women's Canadian Historical Society of Ottawa, Transactions*, XI (1954), 7-16.

Dunham, Mabel. *Grand River*. Toronto, 1945.

Durnford, Mary, ed. *Family Recollections of Lieut. General Elias Walker Durnford*. Montreal, 1863.

Easterbrook, W. T., and Hugh G. J. Aitken. *Canadian Economic History*. Toronto, 1958.

Eccles, W. J., "The Social, Economic and Political Significance of the Military Establishment in New France," *Canadian Historical Review*, LII (March 1971), 1-22.

Edge, R. C. A. "Ordnance Survey at Home." *The Geographical Magazine* (Oct. 1969), 22-23.

England, Robert. "Disbanded and Discharged Soldiers in Canada Prior to 1914." *Canadian Historical Review*, XXVII (1964), 1-18.

ffrench, Yvonne. *The Great Exhibition: 1851*. London, 1951.

Forbes, Archibald. *A History of the Army Ordnance Service*. 2 vols. London, 1929.

Fortescue, J. W. *A History of the British Army*. 13 vols. London, 1899-1930.

Friel, H. J. "The Rideau Canal and the Founding of Ottawa." *Transactions of the Women's Canadian Historical Society of Ottawa*, I (1901), 31-35.

Galbraith, John S. "The 'Turbulent Frontier' as a Factor in British Expansion." *Comparative Studies in Society and History*, II (1959-1960).

————— . *Reluctant Empire. British Policy on the South African Frontier, 1834-1854*. Los Angeles, 1963.

Gallagher, Ruth A. "The Military-Indian Frontier, 1830-1835." *Iowa Journal of History and Politics*, XV (1917), 393-428.

Garwood, F. S. "The Royal Staff Corps, 1800-1837." *Royal Engineers Journal*, LVII (1943), 81-96, 247-60.

Glazebrook, G. P. de T. *A History of Transportation in Canada*. 2 vols. Toronto, rev. ed., 1964.

Glover, Richard. *Peninsular Preparation. The Reform of the British Army, 1795-1809*. Cambridge, 1963.

Goetzmann, William H. *Army Exploration in the American West, 1803-1863*. New Haven, 1959.

Gourlay, J. L. *History of the Ottawa Valley*. Ottawa, 1896.

Graham, Gerald S. "Views of General Murray on the Defence of Upper Canada, 1815." *Canadian Historical Review*, XXXIV (1953), 158-65.

Guedalla, Philip. *The Duke*. London, 1931.

Guillet, Edwin C., ed. *The Valley of the Trent*. Toronto, 1957. Publication of the Champlain Society.

Herrington, Walter S. *History of the County of Lennox and Addington*. Toronto, 1913.

Hill, H. P. "The Construction of the Rideau Canal, 1826-1832." Ontario Historical Society, *Papers and Records*, XXII (1925), 117-24.

————— . "Lieutenant Colonel John By. A Biography." *Royal Engineers Journal*, XLVI (1932), 522-25.

Hind, Edith J. "Troubles of a Canal-Builder: Lieut. Col. John By and the Burgess Accusations." *Ontario History*, LVII (1965), 141-47.

Hitsman, J. Mackay. "Please Send Us a Garrison." *Ontario History*, L (1958), 189-91.

———. *Safeguarding Canada, 1763-1873*. Toronto, 1968.

Hogg, Oliver F. G. *The Royal Arsenal*. 2 vols. London, 1963.

Hunter, A. F. *A History of Simcoe County*. 2 vols. Barrie, Ontario, 1909.

Innis, Harold A. *The Fur Trade in Canada*. Toronto, rev. ed. 1967.

Jackson, W. Turrentine. *Wagon Roads West; A Study of Federal Road Surveys and Construction in the Trans-Mississippi West, 1846-1869*. Berkeley and Los Angeles, 1952.

Jeffries, Charles. *The Colonial Empire and Its Civil Service*. Cambridge, 1938.

Kenny, F. Gertrude. "Some Account of Bytown." *Women's Canadian Historical Society of Ottawa, Transactions,* I (1901), 22-30.

Kingsford, William. *The Canadian Canals: Their History and Cost, With an Enquiry into the Policy Necessary to Advance the Well-Being of the Province*. Toronto, 1865.

Kirby, William. *Annals of Niagara*. Toronto, 1927.

Lajeunesse, Ernest J. *The Windsor Border Region, Canada's Southernmost Frontier; A Collection of Documents*. Toronto, 1960. Publication of the Champlain Society.

Lanctot, Gustave. *Canada and the American Revolution, 1774-1783*. Toronto, 1967.

Landon, Fred. *Western Ontario and the American Frontier*. Toronto, 1941.

———. "British Regiments in London." *Western Ontario Historical Notes,* XIII (Sept. 1955), 1-23.

Leavitt, T. W. H. *History of Leeds and Grenville, Ontario, from 1749 to 1849*. Brockville, 1879.

Legget, Robert. *Rideau Waterway*. Toronto, 1955.

Lehmann, Joseph H. *All Sir Garnet: A Life of Field-Marshal Lord Wolseley*. London, 1964. The title of the Boston edition of the same year is *The Model Major-General: A Biography of Field Marshal Lord Wolseley*.

Macdonald, Norman. *Canada, 1763-1841; Immigration and Settlement; The Administration of the Imperial Land Regulations*. Toronto, 1939.

Macdonnell, J. A. *Sketches Illustrating the Early Settlement and History of Glengarry in Canada*. Montreal, 1893.

Mactaggart, John. *Three Years in Canada: An Account of the Actual State of the Country in 1826-7-8*. 2 vols. London, 1829.

Martell, J. S. "Military Settlements in Nova Scotia After the War of 1812." *Collections of the Nova Scotia Historical Society,* XXIV (1938), 17-105.

Manning, Helen Taft. *British Colonial Government After the American Revolution, 1782-1820*. New Haven, 1933.

Maxwell, Sir Herbert. *The Life of Wellington; The Restoration of the Martial Power of Great Britain*. 2 vols. London, 3rd ed., 1900.

Mealing, S. R. "The Enthusiasms of John Graves Simcoe." Canadian Historical Association, *Report* (1958), 50-62.

Miller, Roscoe R. "The Welland Canal Company and the Duke of Wellington." *Ontario History*, XLVI (1954), 63-67.

Morton, W. L. *The Critical Years; The Union of British North America, 1857-1873*. Toronto, 1964.

Murray, Florence B., ed. and intro. *Muskoka and Haliburton, 1615-1875*. Toronto, 1963.

Ontario, Province of. *The Rideau-Trent-Severn. Canada's unique 425 mile waterway corridor. Yesterday, Today, Tomorrow. A report on optimum recreational development*. Toronto, 1971.

Otway Ruthven, A. J. *A History of Medieval Ireland.* London, 1968.

Parkinson, C. Northcote. *British Intervention in Malaya, 1867-1877.* Singapore, 1960.

Philip, John. "The Economic and Social Effects of the British Garrisons on the Development of Western Upper Canada." Ontario Historical Society, *Papers and Records*, XLI (1949), 37-48.

Phillpotts. George. "Report on the Canal Navigation of the Canadas." Corps of Royal Engineers, *Professional Papers*, V (1842), 145-71.

Piers, Harry. *The Evolution of the Halifax Fortress, 1749-1828.* Halifax, 1947. Publication of the Public Archives of Nova Scotia.

Pike, Douglas, ed. *Australian Dictionary of Biography.* 7 vols. Melbourne, 1966.

Pipes, Richard E. "The Russian Military Colonies, 1810-1831." *The Journal of Modern History*, XXII (Sept. 1950), 205-19.

Playter, George F. "An Account of Three Military Settlements in Eastern Ontario—Perth, Lanark and Richmond, 1815-20." Ontario Historical Society, *Papers and Records*, XX (1922), 98-104.

Porter, Whitworth. *History of the Corps of Royal Engineers.* 2 vols. London, 1889, 1915.

Prucha, Francis Paul. *Broadax and Bayonet. The Role of the United States Army in the Development of the Northwest, 1815-1860.* Madison, Wisc., 1953.

——————— . *The Sword of the Republic. The United States Army on the Frontier, 1783-1846.* New York, 1969.

Pugh, Ralph B. "The Colonial Office, 1801-1925." *The Cambridge History of the British Empire*, III: *The Empire Commonwealth, 1870-1919.* Cambridge, 1959.

Raddall, Thomas H. *Halifax, Warden of the North.* London, 1950.

Raudzens, George K. "A Successful Military Settlement: Earl Grey's Enrolled Pensioners of 1846 in Canada." *Canadian Historical Review*, LII (Dec. 1971), 398-403.

——————— . " 'Red George' Macdonnell, Military Saviour of Upper Canada?" *Ontario History*, LXII (Dec. 1970), 199-212.

——————— "The Military Impact on Canadian Canals 1815-1825." *Canadian Historical Review*, LVI (Sept. 1973), 273-286.

Riddell, W. R. *The Life of John Graves Simcoe.* Toronto, 1926.

Robinson, Ronald, and John Gallagher. *Africa and the Victorians; The Official Mind of Imperialism.* New York, 1967.

Ross, A. H. D. *Ottawa, Past and Present.* Ottawa, 1927.

Roy, James A. *Kingston: The King's Town.* Toronto, 1952.

Rutherford, James. *Sir George Grey, K.C.B., 1812-1898. A Study in Colonial Government.* London, 1961.

Sabine, Edward. *Observations made at the Magnetical and Meteorological Observatory at Toronto in Canada.* London, 1845.

Sage, Walter Noble. "Some Aspects of the Frontier in Canadian History." Canadian Historical Association, *Report* (1928), 162-72.

Salmon, E. T. *Roman Colonization Under the Republic.* London, 1969.

Scadding, Henry. *Toronto the Old.* Toronto, 1873.

Schuyler, R. L. "The Recall of the Legions: A Phase of the Decentralization of the British Empire." *American Historical Review*, XXVI (1920), 18-36.

Scott, R. W. *Recollections of Bytown.* Ottawa, 1908.

Simpson, J. R. "Some Reminiscences of Bytown." *Transactions of the Women's Canadian Historical Society of Ottawa*, VII (1917), 5-11.

Skentelbury, N. *The Ordnance Board; An Historical Note.* War Office Library Papers, Annex A to Proc. 40, 169, Sept. 1964.

Smyth, Sir James Carmichael. *Precis of the Wars in Canada.* London, 1862.

Sneyd, Robert Brown. "The Role of the Rideau Waterway, 1826-1856." M.A. Thesis, University of Toronto, 1965.

Stacey, C. P. "An American Plan for a Canadian Campaign. Secretary James Monroe to Major General Jacob Brown, February 1815." *American Historical Review*, XLLI (1941), 384-358.

——— . "Halifax as an International Strategic Factor, 1749-1949." Canadian Historical Association, *Report* (1949), 46-55.

——— . "The Backbone of Canada." Canadian Historical Association, *Report* (1953), 1-13.

——— . *The Undefended Border; The Myth and the Reality.* Ottawa, 1962. Canadian Historical Association pamphlet no. 1.

——— . *Canada and the British Army, 1846-1871; A Study in the Practices of Responsible Government.* Toronto, rev. ed., 1963.

Stephen, Leslie, and Sidney Lee, eds. *The Dictionary of National Biography.* Oxford, 1917- .

Stewart, McLeod. *First Half Century of Ottawa.* Ottawa, 1910.

Taylor, George Rogers. *The Transportation Revolution, 1815-1860.* New York, 1951.

Thomson, Don W. *Men and Meridians; The History of Surveying and Mapping in Canada.* 2 vols. Ottawa, 1966.

Twitchett, Denis. "Lands Under State Cultivation Under the T'ang." *Journal of the Economic and Social History of the Orient*, II (1959), 162-203.

Veitch, David. "The Royal Engineers and British Columbia." *Royal Engineers Journal*, LXXII (1958), 106-17.

Vernadsky, George. *A History of Russia.* New Haven, 1969.

Ward, S. P. G. *Wellington.* London, 1963.

Watson, G. R. *The Roman Soldier.* Bristol, 1969.

Winks, Robin W., ed. *The Historiography of the British Empire—Commonwealth; Trends, Interpretations, and Resources.* Durham, N.C., 1966.

Wesley, Edgar Bruce. *Guarding the Frontier; A Study of Frontier Defense from 1815 to 1825.* Minneapolis, 1935.

Wise, S. F. "The Indian Diplomacy of John Graves Simcoe." Canadian Historical Association, *Report* (1953), 36-44.

Woodward, Sir Llewellyn. *The Age of Reform, 1815-1870.* Oxford, 1962. Oxford History of England Series.

Wraith, R. E. *Guggisberg.* London, 1967.

Wrottesley, George. *Life and Times of Field Marshal Sir John Fox Burgoyne, Bart.* London, 1873.

A Technical Assessment of the Rideau Canal from the Commencement of Construction

The only contemporary evaluation of the Rideau Canal project as a whole from the technical point of view was written by one of Colonel By's officers, Lieutenant E. C. Frome, R.E. (see Chap. II, note 25, and p. 17). His paper, "Account of the Causes which led to the Construction of the Rideau Canal, connecting the Waters of Lake Ontario and the Ottawa; the Nature of the Communication prior to 1827; and a Description of the Works by means of which it is converted into a Steam-boat Navigation," published in *Papers and Subjects Connected with the Duties of the Corps of Royal Engineers* (London, 1837), is reprinted in full.

DURING the late war with the United States, the communication between the Upper and Lower Provinces of Canada, was liable to constant interruption, from its proximity to the American frontier for at least two-thirds of the distance between Montreal and Kingston, which latter town, situated at the eastern extremity of Lake Ontario, and the source of the St. Lawrence, being the principal depôt for naval and military stores intended for the upper province, and containing a large dock-yard, where all the vessels for the defence of the lake are fitted out, becomes in time of war a place of the first importance, and is in fact the key to Upper Canada. Independent of the risk of interruption from an enemy, the difficulty and expense * of moving heavy military stores and

---

* The cost of the transport of merchandise by the St. Lawrence, from the head of the island of Montreal to Kingston, was in 1815 from 4*l*. to 4*l*. 10*s*. per ton. The Durham boats used for

ammunition at all seasons along the vile roads by the side of the rapids on the St. Lawrence was excessive; and had the Americans made a more vigorous use of their powers of annoyance, the transport of the two frigates sent out in pieces from England, with all their guns, cables, and ammunition, by this route, would have been quite impossible.

These considerations showed the obvious advantage, and in the event of another war, the absolute necessity, of some more secure communication with Kingston, which would be beyond the reach of an enemy, and by avoiding any land-carriage or transhipment of stores, would materially diminish the expense of transport. A reference to the map of Upper Canada will show, that the route by the Ottawa and Rideau rivers can alone combine these desiderata, and at the close of the war, in 1814, the attention of Government was turned to this part of the country, of which little was at that time known.

In the spring of 1815, Colonel Nicolls, then commanding Engineer in Canada, was directed to send an officer of Engineers to explore and report upon it generally; and Captain (then Lieutenant) Jebb, was instructed to proceed upon this duty. His orders were to follow up the course of the Cataroqui from Kingston Mills, and keeping a northerly direction, to penetrate into the Rideau Lake, and descend the river which flows from it to its confluence with the Ottawa. He was to return up the river as far as the mouth of the Irish creek, and trace the waters of which it is the outlet to their source, and from thence to follow up the best communication he could find to Kingston Mills, or to Gannanoqui, and suggest any temporary expedients for improving the navigation, so as to render it available for batteaux. He was also directed to note the nature of the country, with a view to its being deemed eligible or otherwise for the establishment of military settlements. Captain Jebb stated on his return, that he considered either of these routes might be rendered practicable, but gave the preference to the shortest by the Irish Creek. The works he proposed were, a dam at the mouth of the Cranberry Marsh, and another where the Ganna-noqui leaves the White Fish Lake; the Irish Lake to be deepened by a dam at the mouth of the creek, and the channels of some of the rapids on the Rideau to be cleared. All these suggestions however were of such a nature as

---

this purpose require six men, and take on an average twelve days in ascending, though only four in returning, and carry about eight tons.

The batteaux also used on the river require fewer hands, and hold about five tons.

he considered would come within the scope of the very limited sum proposed to be expended.

Nothing was undertaken in consequence of this report, but the settlements of Perth, and some years afterwards of Richmond, on two small rivers flowing into the Rideau, were laid out, and the town lots granted mostly to officers and men of regiments disbanded in Canada after the war.

In 1821 the subject was taken up by the provincial government of Upper Canada, who passed an act " to make provision for the improvement of the internal navigation of the province," and appointed commissioners to determine upon the practicability, and report upon the estimated expense, of such undertakings as they might propose; on which they were occupied for nearly four years. The falls of Niagara, and the rapids of the " Long Sault" on the St. Lawrence, being the two principal obstacles to an uninterrupted communication between Lake Erie and the lower province, their attention was naturally first turned to these points,* and in 1823 they sent in estimates for a canal to connect Lake Erie and Burlington Bay, and also a cut between Burlington and Lake Ontario. The following year produced three estimates for canals, of different sized locks, and depths of water, connecting Lake Ontario and the Ottawa by the Rideau, and also the Bay of Quinté and Presqu' Ile Harbour; all the details of which were published by the joint committee of the two houses of legislature, to whom the reports were referred in 1825. Previous to this, the Welland canal had been undertaken by a chartered company, leaving only the obstacles on the St. Lawrence to be overcome by improving the navigation of the river itself, or by connecting Lake Ontario and the Ottawa, either by the proposed route of the Rideau; the Mississippi, which joins the Ottawa above the Chaudiere Falls; or the Petite Nation, leaving the St. Lawrence a little above Prescott. In a commercial point of view, the first and last of these were considered perhaps preferable, but they would both have been exposed to interruption along the line of the frontier. An estimate was however framed for the improvement of the St. Lawrence as far as the boundary of the Upper Province, and application was made to the provincial government of Lower Canada, to carry the same down to Montreal, and assist in devising some

---

* The La Chine canal, to avoid the rapids opposite Montreal, was already commenced by the Lower Province, and that at Grenville to turn the " Long Sault " on the Ottawa, by the British Government.

method of avoiding the interference likely to be met with on account of the cession of Barnhart's Island to the Americans. The sum of 5000*l.* which had accumulated during some years from tolls levied on all rafts passing Chateaugay, was recommended by the arbitrators named by the two provinces to be applied to the purposes of this survey. Of the two estimates made by the Upper Province, the first amounted to 176,378*l.* for locks 132 feet by forty, with eight feet water; the second to 92,834*l.* for locks of 100 feet by fifteen, and four feet water.

The estimates for a water communication by the Rideau, were for the following descriptions of canals, by the line of the Rideau Lake, that by the Irish Creek, though the most direct, having been abandoned by Mr. Clowes, the engineer employed by the province, on account of the comparative scarcity of water, and the depth of cutting at the summit level, neither of which objections appear to exist to the extent stated by him.—First estimate, canal seven feet in depth, forty feet in width at the bottom, and sixty-one at the surface of the water, the banks to slope one and a half to one; the locks to be of stone 100 long by 22 wide; cost 230,785*l.* Second, five feet in depth, twenty-eight feet at bottom, and forty-eight on surface; locks of stone eighty feet long by fifteen wide; cost 145,802*l.* Third, four feet water, twenty feet wide at bottom, and thirty-two on the surface; locks of wood seventy-five feet long by ten wide; cost 62,258*l.*

Of these projects the committee appeared to give the preference to the second, but looking forward to the promised loan of 70,000*l.* offered by the British Government, (if they did not expect even more substantial assistance,) they add in their report, "that if the parent State should assume a share of the charge, it would be highly expedient to attempt a canal of the larger dimensions."

As the circuitous route by which it was proposed to open a secure water-communication between the lakes and the lower province, owed its adoption solely to its being considered essential to the future defence of the country; the British Government could not be blind to the advantages of retaining in their own hands, the complete control over the work, in case of another war; and in 1825 a committee, consisting of Colonel Sir J. C. Smith, Lieutenant-Colonel Sir G. Hoste, and Major Harris, Royal Engineers, sent out from England to report upon various public works in Canada, were directed to bring home an estimate for the Rideau canal, the locks to be on the same scale as those of La Chine and Grenville, 108 feet long and twenty broad, and the depths of all the necessary excavations five feet.

The amount of this estimate was 169,000*l.*, exceeding the second of Mr. Clowes (on whom they relied for their details and levels) by 20,000*l.*, on account of the increased size of the locks.

In consequence of this report, its construction was determined upon, the expense to be borne wholly by the British Government; and in May 1826, Lieutenant-Colonel By, Royal Engineers, was sent out to superintend the works; the sum of 5000*l.* having been granted him for the first year's expenditure, principally for the purpose of exploring and clearing the country, and obtaining such local information as would enable him to commence operations the moment means were furnished him. His further instructions for the progress of the work could not, from its nature, and the little that was known of the difficulties to be encountered, he very closely defined; but as a general standard for the size and construction of the locks, he was referred to the La Chine canal, and that in progress at Grenville, by the Royal Staff corps; and as at this period it was not contemplated as a steam-boat navigation, towing-paths were to be formed wherever practicable.

Colonel By reached Montreal in June, and in September 1826, proceeded up the Ottawa to the mouth of the Rideau, where, according to Mr. Clowes' plan, the canal was to join the Ottawa. He however decided upon a deep bav about a mile above the former spot, as better adapted for the entrance; and the excavation for the locks was commenced immediately, as also a series of bridges across the Ottawa at the Chaudiere Falls (about half a mile higher up), connecting the two provinces, and opening a communication with a small tract of cleared country at Hull, on the lower Canadian side. Some improvements were likewise made in the timber channel for rafts at these falls; for which service, as totally unconnected with the canal, the sum of 2000*l.* was authorized by Lord Dalhousie to be expended from the military chest.*

The following spring Colonel By took up his residence at the entrance of the canal, which was considered during the execution of the work as head-quarters, and barracks and a hospital were erected there for two companies of sappers, which were sent out from England for this service. The works at the different stations were many of them given out to contractors during the

---

* The largest of the wooden truss bridges, 212 feet span, fell during its construction, just as the string-pieces were placed across the chasm; it was reconstructed and stood till last year, when it again gave way, as well as the smaller one on the same principle. The cause of both failures was the same; the want of proper abutments.

early part of the summer, when Colonel By went through the intended line to Kingston; and in the months of July and August the necessary details were furnished by an officer of Engineers, sent with a party for the purpose, upon which a report and estimate was framed, and sent to England in the autumn of 1827.

The amount of this estimate was 474,000*l.* for a water-communication, similar in size and construction of the locks to the La Chine canal, and the work was at once commenced, adhering in many instances to the plan recommended in the report, but in others deviating in part, or even totally from it, as the further development of the ground, when the land became cleared, rendered advisable. It is therefore useless to dwell upon the suggestions contained in it; but before proceeding to any account of the works, as at present finished, it is necessary to describe the nature of the communication as it existed prior to the spring of 1827, omitting all notice of the line by the Irish Creek, which was at once abandoned without meeting with the consideration it deserved, on account of the report given of it by Mr. Clowes.

The river Rideau falls into the Ottawa about sixty miles above the head of the canal at Grenville, over a ledge of rocks more than a mile below the present entrance of the canal. This fall varies from twenty to upwards of thirty feet, according to the season of the year, as the two rivers are not at their height at the same time, the freshes on the Rideau having subsided, before the principal rise of the Ottawa, occasioned by the melting of the snow in the distant mountains of the north, has commenced. The river for some miles above its junction is almost a continued rapid, and is so obstructed that large canoes frequently made a portage to the *Hog's Back*, a distance of five miles, in preference to poling up the stream. The level of the still water at this place was nearly 115 feet above the Ottawa, and it continued for about four miles to the *Black Rapids*, where was a fall of nine feet in a few hundred yards. Another sheet of still water extended five miles and a half to the foot of *Long Island*, just below which the river Jack, or Goodwood, flowing through the military settlement of Richmond already alluded to, empties itself into the Rideau; and though in the summer the water over the rocks at its junction was hardly six inches deep, in the spring it swelled to a considerable stream, down which a quantity of timber was rafted. The two narrow channels into which this island divided the river were rapids for nearly their whole length of three miles, which generally required (to use the term of the country) a "portage," the whole fall being nearly twenty-five feet. At its foot was a small clearance with a paltry saw-mill on the right bank of the river, but the island and the banks were rocky and uncultivated. From its head was a continued sheet of still water for twenty-three miles, the banks on each side being low, marshy, and totally uncleared, excepting two or three small farms near Garlick's on the left bank, where Stephen's Creek empties itself. Within five or six miles of the end of this sheet of water, is the mouth of the South branch, three miles up which stream was a settlement and mill, known by the name of the owner, "Chetham." For seven miles higher up the Rideau, the country is of a better description, and was settled about thirty years before the commencement of the canal, principally by emigrants from the United States. In this distance the following impediments were met with in the river; "*Burritt's Rapids*" and *Hurd's Shallows*, immediately above the still water, at each of which a wooden bridge crossed the river; *Nicholson's Rapids* two miles higher, and *Merrick's Rapids* and *Falls*. The total rise by all these being nearly sixty feet. At the last named place a rough dam had been formed across the river, with a grist and saw mill. There was a tolerable road to Brockville, distant thirty-two miles, and another to Prescott five miles nearer. The clearance terminated half a mile above this dam, and another eight miles of still water, with low swampy banks, extended to *Maitland's Rapids*, where a road from Brockville crossed the river by a ford. * Another interval of still water, for four miles, to *Edmons,'* was succeeded by a series of small rapids for nearly two miles, to *Old Slys*, and about a mile higher up, the river rushed down a descent of about thirty-five feet in a few hundred yards, over a rough mass of limestone rock. The channel was not much above fifty feet wide, and advantage had been taken of this fall to establish a saw-mill and dam by a settler, from whom the place received its name of *Smith's Falls*, but there was no clearance of any size. For nearly three miles above this mill, a series of small rapids extended to what was termed the *First Rapids* on the Rideau. A stream called Owl Creek entered the river at the lower end of this distance, and Cockburn Creek just above the first rapids; the latter has its source near the Mississippi river, and may hereafter be the means of connecting its waters with the Rideau canal.

---

* A mile and a half below this is the mouth of the Irish Creek, once proposed as the line of the intended communication; and about the same distance, still lower, a large stream, called Barber's Creek, enters the Rideau, both on the right hand.

The entrance to the *Rideau lake* may be said to commence two or three miles above this creek, and about another mile further up, the river Tay empties itself into the lake. The town of Perth, the largest of the two military settlements, is situated on this river, eleven miles from its mouth, and its inhabitants have, since the commencement of the Rideau canal, formed a navigation for small steamboats as high as the town, by a series of dams and locks at the different rapids.

The direction of the length of the Rideau lake is nearly south-west, and the distance from the first rapids on the Rideau, to the isthmus where the line of canal leaves the lake, is twenty-three miles. At *Oliver's Ferry*, where the road from Brockville to Perth crosses, it contracts to 400 yards, and opens out again in a deep bay to the eastward, making the extreme breadth of the lake, which is full of small islands, about seven miles. At the *Upper Narrows* it was again contracted to a channel of not more than 100 feet, by a narrow low tongue of land which projected from the north-western shore, and has been since cut through. This channel was three miles from the *Isthmus,* between the Rideau and Mud lakes, where canoes, (the only mode of conveyance that can be used on the small rivers and isolated lakes of Canada,) were obliged to make a " portage" of a mile and a half. The level of the *Mud lake* was three feet and a half below that of the Rideau, and its shores were rocky and uncleared. Crossing it in the direction shown in the sketch, a narrow passage, through bold rocks of granite, led into *Clear Lake*, between which and *Indian lake*, a strip of land about 180 feet broad, required another portage. A winding creek led from this to *Chaffey's Mills*, where a dam had been formed to obtain a head of water to work a saw-mill and distillery, to supply the few scattered settlements in the neighbourhood ; the shores of all these lakes being uncleared, and, from the rocky nature of the soil, of little value. The fall at the mill-dam was about thirteen feet, and its distance from the entrance into Mud lake was five miles. Following the creek below the dam, and keeping the eastern shore of a small lake (called *Musquito*, or *Opinicon*) led into another channel, where was also a wooden dam and saw-mill, with a fall of about seven feet and a half, belonging to a settler of the name of Davis. The outlet of the small lake below this, (*Sand Lake*), was through a narrow crooked ravine, with high rocky banks of sandstone and a species of granite. By this channel the waters of these lakes emptied themselves into *Cranberry Lake*, falling about sixty feet in a distance of less than a mile. For ten miles below this, the route continued through the lake, and by a channel through a quantity of marshy land which had been flooded by dams erected at the *White Fish* falls, and at the *Round Tail*, the source of the *Cataroqui*.*

A portage was made round this dam, where was a fall of five feet, and another a mile below, where an establishment, consisting of saw and grist mills, had been erected by a Mr. Brewer, with a fall of water of eleven feet. The banks, on first entering this river, were rocky and precipitous, but they rapidly changed their character after leaving *Brewer's Upper Mills*. Two miles below was another dam and mill, belonging to the same individual, with a head of about eleven feet. Two small rapids only occurred between this place and Kingston mills, which in the spring were hardly discernible, the narrow sluggish stream winding its tortuous course between clay banks not more than three or four feet above its surface, and occasionally passing large swamps with a quantity of dead timber standing on the flooded land ; its breadth did not average more than sixty feet, and though generally deep enough for boats, particularly in the spring, at the end of summer it was in many places too shallow even for loaded canoes.

At *Kingston Mills* a saw-mill had been erected, at the expense of Government, with a rough sort of dam abutting upon the granite rocks, which here contracted the channel. The fall from the mill-pond was about twenty-six feet, into the head of a deeply-indented bay of Lake Ontario, about five miles from the town of Kingston. This bay, generally known by the name of Cataroqui, was nearly choked up with sedgy islands, between which a channel was kept clear by the stream, but about a mile from the mill-dam a rocky bar crossed it, on which there was sometimes not more than three feet of water.

From this description of the chain of lakes and the rivers that were to form this communication, it appears obvious that the general features of any work, to render them navigable, must consist of a series of dams and locks at each obstruction, raising the water at the foot of each rapid to the level of that above, where practicable, and locking by the side of the dam. The excavation necessary for canals in a cultivated country is thus saved, and the rivers and lakes converted, from a succession of falls and rapids, into steps of still water at different levels. This was the view taken by Lieutenant-Colonel By, and acted upon by him perhaps in excess ; and the same plan was also adopted, though

---

* The creek across which this dam was erected, leads into the White Fish Lake, from whence flows the river Gannanoqui, entering the St. Lawrence about eighteen miles below Kingston.

not to the same extent, by Mr. Clowes, the engineer, previously employed by the province. Canals in a country where water is of value and scarce, are of course entirely different in their principle from those suited to Canada; where the only difficulty consists in attaining a complete control over the immense quantity of water passing through a great extent of partially explored country, particularly during the melting of the snow and ice in the spring.

The enlargement of the locks from their original plan was authorised by the committee, consisting of Sir J. Kempt (then Lieutenant-Governor of Nova Scotia), and Colonels Fanshawe and Lewis, Royal Engineers, who were ordered, in the spring of 1828, to examine into the details of the Rideau canal on the spot, and decide upon several alterations which had been suggested, and referred to a committee of Engineer Officers in Pall-Mall, of which the late Sir A. Bryce was President, in January 1828. Such portions as were necessary of those which had been commenced on the smaller scale were pulled down, the contractor being paid for the work done by measurement, and afterwards allowed the materials for the trouble and expense of removing them.

The following extract from their Report shows the general state in which they found the existing navigation of the Ottawa and Rideau between Montreal and Kingston.

" This line may be divided into two portions: 128 miles from Montreal by the Ottawa to By-Town at the entrance of the Rideau; and 154 miles by the Rideau river and lake navigation from By-Town to Kingston.

" The first commences with the La Chine canal at Montreal, extending nine miles, and is complete for vessels not exceeding twenty-feet beam, and five feet depth of water. This is a provincial work.

" At the junction of the Ottawa and St. Lawrence, at the western extremity of Montreal Island, are St. Ann's Rapids and Vaudreuil Passage, separated by the Isle de Perrot, and not navigable for vessels of the above draft in dry seasons. But from hence, for a distance of twenty-seven miles to the foot of Carillon Rapids, such vessels may be used. The Carillon Rapids are not navigable, and will require a canal one mile and a half long with two locks.

" The Chûte à Blondeau is a short rapid, intermediate between the Carillon and Grenville, which will also require a canal and one lock. At the foot of the Long Sault of the Ottawa, commences the Grenville canal, now executing by the Royal Staff corps, originally intended for vessels of twenty-feet beam, and four-feet draft of water, but will admit of six-feet water. Three out of the six locks intended for the Grenville canal, and nearly all its excavations, are completed. This distance of interrupted navigation, from the foot of the Carillon to the head of the Long Sault, is about twelve miles ; from hence to the entrance of the Rideau navigation is a distance of forty-four miles, having occasional shoals, with not more than five-feet water in dry seasons."

The progress then made in the Rideau navigation is next dwelt upon, but it is unnecessary to allude to it here.

The Carillon, Chûte à Blondeau, and Grenville canals have since been completed, and such of the locks as were not finished, or nearly so at this period, have been built of the enlarged size; the others have not been altered, and prevent the steam-boats which ply on the Rideau from passing through the Grenville canal.

The whole chain of communication between Montreal and Lake Erie is now accomplished by seven canals, of which the Rideau is but one section. They have been already alluded to, but are given below in the order in which they occur.

1st. The Welland canal, uniting Lakes Erie and Ontario.

2nd. The Rideau, connecting the waters of Lake Ontario and the Ottawa.

3rd. The Grenville canal, turning the rapids of the " Long Sault" on the Ottawa.

4th. The Chûte à Blondeau, consisting of a single lock and canal, less than a mile long.

5th. The Carrillon, of two locks and a canal, about two miles and a half long. The three last are generally known as the " Ottawa Canals," the distance, from the head of the Grenville to the foot of the Carillon, being about twelve miles. They were all constructed by the Royal Staff corps, and were commenced in 1821 and finished in 1834.

6th. A single lock at the " Vaudreuil" passage, thirty miles below the Carillon, built by and in possession of a private company.

7th. The Lachine canal, extending nine miles from the head of the rapids to the town of Montreal. This is a provincial work, but the Government have a right to a free passage for all stores and troops, having advanced 12,000*l.* during its construction on that condition.

When the projected canal at St. Ann's is constructed, the lock at the Vaudreuil passage will become useless, and should the rapids on the channel at the back of the island of Montreal be hereafter rendered navigable, the communication will be unbroken between the Atlantic and the Upper Lakes. At present it is interrupted at Montreal, and the interests of the merchants of that town, are of course opposed to any plan by which it would be avoided.

The accompanying map shows, by the dotted line, the present route for steam-boats between Montreal and Kingston.

Plate 2, fig. 1.—To return to the subject of the works by which the part of this route between the Ottawa and Lake Ontario has been converted into a steam-boat navigation. At the entrance from the Ottawa, eight locks, built on invert arches, are placed in succession, the breastwork of the river lock (allowing seven feet water over the floor), being eleven feet high, and each of the others ten feet. There being also seven feet on the upper sill of the upper lock, the total lift is eighty-one feet from the surface of the average lowest water in the Ottawa. The soil in which the lock-pits were excavated, is rather a stiff clay, mixed with a few boulders, and loose veins of sand were met with near the centre. Immediately above the locks, the canal for a short distance is cut through rock. The material of which the locks are built is a compact limestone, and was quarried on the cliffs on each side of the excavation. Nearly the same species of stone is used for all the locks as far as the Rideau Lake. Two large stone buildings were erected at an early stage of the work on each side of the locks for a commissariat store, and Engineer office ; and barracks and a hospital, on the hill above the entrance, for the two companies of sappers, who were for the first summer encamped on the opposite height, the country being at that time a perfect wilderness. Work-shops were also built on a large scale, in which the artificers from the sappers were employed on the lock-gates, &c. as well as a number of civil carpenters and blacksmiths ; and a large rambling village soon sprung up at this spot, containing a number of stores to supply the wants of the men employed at the works immediately in the neighbourhood. A stone bridge crosses the canal above the upper lock, and connects the buildings on each side of it.

These locks were among the first that were tried, and the water forced its way through the breastworks ; in many instances moving the large stones which formed the sills. These were afterwards secured with a number of fox-wedge bolts, five or six feet long, and heavy iron straps connecting them at the angle. A quantity of cement, of which none had been used in building, was forced into the breasts, side-walls, and floors, in the shape of grout, by means of long tin tubes ; and being allowed time to consolidate, it has rendered them nearly water-tight. This same expedient has been followed in almost all the works, particularly those which were not founded on rock, and has every where been found to answer. The cement was made from a stone quarried on the opposite side of the Ottawa, which, being burnt and ground very fine, proved a better water-cement than some obtained from the States, and far superior to the Harwich cement, which was nearly spoilt before it reached the canal.

Above the head of the first eight locks is a basin surrounded with an earthen embankment, from whence a channel is cut, with a pair of floodgates at its entrance, by means of which the canal may be drained. The *deep cut* extends in the same line as the locks, running east for three quarters of a mile, till it enters a natural ravine, whose course is about north and south, the average cutting being twenty-five feet. The soil is a stiff clay, but which is very soluble in water, and occasioned considerable trouble and expense in excavating, having slid more than once " en masse " into the canal.

An embankment across the northern end of this ravine where it meets the deep cut, retains more than the required depth of water for nearly two miles to the *Notch of the Mountain*, where the cutting recommences, and continues, at an average depth of twenty feet, through gravel and boulders for half a mile, to *Dow's Great Swamp*, which by means of two massive earthern embankments, is converted into a pool twenty feet deep.

About seven hundred yards more excavation through a swamp, at the average depth of three feet, brings the canal to *Hartwells*, where are two combined locks built of limestone upon invert arches, on a foundation of clay mixed with large boulders, and full of springs. Blocks of rough stone were thrown in between these boulders, and the masonry of the locks built upon them, no piles having been used. In the left wing of the upper lock a small regulating sluice is constructed, capable of emptying the canal above if required, part of its floor, which is of wood, being one foot below the level of the bottom. An immense quantity of cement grout was forced into the masonry, principally of the invert arches; holes being drilled into the work at intervals, and the short end of the bent tube inserted some inches, and well seamed with clay. A large funnel was formed at the other end, and the liquid grout poured in had thus a pressure of twelve or fifteen feet, according to the length of the tube.

Between this work and the *Hog's Back* is one mile of cutting, running on the slope of the left bank of the Rideau for nearly half the distance. Below the locks at both places, the steep slopes were reveted with rough stone, to prevent their being injured by the rush of water from the sluices.

At the *Hog's Back*, a little more than four miles from the entrance, the canal first enters the Rideau on its left bank. Of the two combined locks constructed here, the upper is only a guard-lock, its coping being eight feet above the surface of the seven-feet water on its sill; the lift of the other is fourteen feet six inches, allowing six feet in the lock; and the height of the masonry of the breast is thirteen feet six inches. The stone used in building was quarried on the opposite side of the river, and is very similar to that at the works below; the excavation was also clay mixed with boulders, and the walls and invert arches were built upon large rough stones without any piling. The recesses were planked.

The dam, one end of which abuts against the wing-wall and backing of the upper lock, was intended to have been of the construction originally proposed in the general specification; the masonry of large stones placed on edge, each alternate one of different height, so as to break joint, with a thickness of puddle behind this wall, and the remainder of the mass and the slope up-stream of clay and rubbish; the water flowing over the top. The first attempt made by the contractor was destroyed early in the spring of 1828, after the arched key-work (as the curved masonry was called), had been partially raised to the height of thirty-seven feet, by the water turning the flank of the dam on the side of the locks, and carrying away a large mass of the clay bank on which it abutted. Soon after this, the contractor gave up the work; and the two companies of sappers, with some hundred labourers, were employed upon it all the winter of 1828 and 1829. The masonry was nearly completed, more than twenty-five feet thick at the base, before the breaking up of the frost; but on the 6th of April, the water forcing its way through the mass of frozen earth which had taken all the winter to accumulate, made an enormous breach near the centre of the dam, carrying away " en masse " every thing opposed to it. It was afterwards finished under the superintendence of Captain Victor, Royal Engineers, by forming a strong frame-work of timber in front of the breach, which was afterwards filled with stone, and supported in the rear by a quantity of large stones thrown in from the top, and filled in front with an enormous mass of clay, stone, and gravel; the base, extending about 250 feet up the stream, forming a slope of about five to one. It is now one of the most substantial works on the whole line of the canal.

Luckily the rock, which on the left bank of the river, (where the locks are placed), only rose twelve or fourteen feet above the bottom, on the opposite side was nearly forty feet, and made an excellent floor for the channel cut to carry off the surplus water. The waste-weir is framed of timber, strongly bolted to the rock, and backed with large blocks of stone. In spring-floods this helps to carry off the rush of water, but the main channel is sufficient of itself at other times. The top of the dam is on the same level with the coping of the lock, eight feet above the average surface water. It forms a sheet of still-water to the foot of the works at *Black Rapids*, a distance of four miles.

Plate 2, fig. 2.—This work consists of a single lock on the left bank of the river, built on a rock foundation, and a dam about twelve feet high, constructed on the original plan, with the water flowing over it; but, to avoid the injury likely to be sustained by the large volume in the spring, a cut-stone sluice-way was formed on the bed of the river, between the wing-wall of the lock and the dam, closed as required with squared logs lowered down a groove in the piers of the masonry. Even the small quantity of water that passed over the dam after this sluice was constructed, carrying a mass of floating timber down with it, caused considerable injury to the rock-foundation upon which it was built, and to the dam itself. The lift of the lock is nine feet, allowing seven-feet water on the lower, and six feet on the upper cill. The height of the breast-work being ten feet.

Plate 2, fig. 3.—At the foot of *Long Island*, five miles and a half higher up, are three combined locks on the right bank of the river, on a rock foundation; and a dam thirty-one feet, on the original plan; but from the damage done at the work below, by allowing part of the water to flow over a height of only twelve feet, it was not deemed prudent to try the experiment here; and a large channel was cut through the clay, on the opposite side of the river to a ravine, through which the surplus water finds its way back into the river about a quarter of a mile below. The dam was raised four feet, to guard against the spring-floods, and an embankment between it and the sluice-way continued on the same level. This sluice-way consisted of three channels of thirty-three feet wide, between piers of cut stone sixty feet long, the centre about six feet deeper than the two extreme ones. The floors were of timber, continued for some distance below the end of the piers, but where this flooring ceased, the water formed deep pools, under-mining the whole work, which has since been totally destroyed, and a year ago, a new sluice-way was constructed on a better principle. Another channel, to carry off a portion of the water, was also formed by taking advantage of a creek a mile and a half up the river, and leading it into the same ravine. The total lift of the three locks is twenty-five feet three inches, allowing eight feet five inches water in the lower lock, and seven feet against the upper gates.

Plate 2, fig. 4.—At *Burritt's Rapids* (twenty-five miles and three quarters above), is a single lock and dam ; the former built on the right bank of the river on a very stiff clay foundation, with a wooden floor ; the lift is ten feet, seven-feet depth of water being on the upper sill, and nine feet in the lock, the bottom of which is at least three feet lower than a rocky shoal in the bed of the river a short distance below. The canal above is about a mile and a quarter in length, and advantage having been taken of a hollow into which water always flowed in the spring, the cutting was not deep, but near the lock the embankment is considerable. A permanent wooden bridge crosses the canal, about 900 yards above the lock, where a road led to a bridge across the river. The lower string-pieces are twenty-eight feet above the surface-water, to allow a sufficient height for the chimnies of steam-boats. The upper sill of the lock is one foot two inches lower than the floor of that of the work above ; but the canal is not excavated quite so deep, part of it being through rock. The dam could not be constructed at the foot of the small rapids, on account of the flat country on the opposite shore, but is situated within less than 400 yards of the head of the canal : it is formed of timber framed and bolted to the smooth rock, and the intervals between the bays are filled up with pine-logs squared at three sides to fit the uprights, and to be nearly water-tight when laid one upon another. The upper row can be removed, to prevent any great rise of water in the spring, and several whole bays taken up if necessary. The average height is about eight feet, and the length between eighty and ninety yards, each end abutting upon a cut-stone pier.

Plate 2, fig. 5.—At the next work, *Nicholson's Rapids*, distant one mile and three quarters, five feet ten inches water is backed up by this dam into the lower lock, which is situated within 200 yards of the river; a canal nearly 400 yards long intervenes between it and the other lock, with a rough stone embankment backed with clay on the river side, the other cutting into rather a steep bank. Above this a similar canal extends about the same distance till it enters the river, the excavation being altogether 1120 yards long, and entirely through rock, except a small portion near the river below. Close to the head of the canal a regulating sluice-way of timber, between rough stone piers, is formed in an opening cut through the rocky bank to the river, to prevent any very large body of water flowing over the dam, which is formed on a plan something similar to that at Black Rapids, but instead of clay, the slope up-stream consists almost entirely of gravel and broken stones ; the section is much diminished by this change, and there is no risk of its washing away, and the puddle sinking at the back of the key-work ; the height averages about nine feet, and to prevent the rock being worn away by the fall of water, two steps of large rough stones are laid to break its fall. The lift by these two locks, which are founded on remarkably level beds of limestone, is fourteen feet ten inches, seven feet six inches by the upper, and seven feet four inches by the lower, reckoning upon five feet six inches water in the canal above. Their breastworks are seven feet and eight feet two inches.

On entering the river, boats cross to the opposite side, where the next lock is entered only 300 yards above. It would have been a far better plan to have kept the right bank, and built the lock there, but it was nearly finished before the lower work was laid out, which had to be altered, from its original impracticable plan of a single lock and dam, on account of an error in the levels.

Plate 2, fig. 5.—The lock at this place, " *Clowes' Quarry*," has no breastwork, the whole lift of nine feet six inches being thrown on the upper gates, and the coping of the walls having a height of five feet eight inches over the surface-water. The excavation for the lock-pit, and the short cut of about 130 yards below it, was through a loose shelly limestone, which forms the floor of the lock ; both the sills are of wood, bolted to the rock, and nearly on the same level. The grooves in the upper-wing walls are filled up with stop-gate logs permanently wedged in to a height that allows between five and six feet water over them ; and the few inches interval between them, and the broken face of the rock that was excavated for the breast, filled up with small stones laid in cement. The dam is about fifteen feet high, and 300 water-way (the wings being raised to the level of the coping of the lock), was commenced on the original plan, but the puddle which washed through the stones and left the key-work in many places bare for a depth of several feet, was replaced with broken stone. Between the dam and lock, a cut-stone regulating sluice, with grooves for squared logs, is formed similar to that at the Black Rapids, the piers being founded on the rock forming the bed of the river, and raised as high as the coping of the lock, but its direction is turned towards the centre of the curve of the dam, to save the backing of the lock from injury. Two steps of large stone are laid close to the base of the key-work of the dam, to save the rock, which was considerably worn, the first spring that the water was allowed to flow over.

Plate 2, fig. 6.— At the old settlement of *Merrick's Mills,* two miles and a half higher up, the works are all on the right bank of the river, and are totally different from those first proposed, where the locks were to have been built in a channel which had been cut round the old saw-mill dam for rafting timber. The embankment necessary at the head of the locks, had this project been attempted, would have been a much more serious undertaking than was anticipated when the estimate was made in 1827.

The three locks are now all detached, the lowest, into which seven feet three inches is backed by the dam below, being within 150 yards of the river. The excavation, excepting this distance, is mostly through a shelly sort of lime-stone, and would average about the depth necessary for the canal; the gradual slope of the land allowing the locks to be placed so as to balance the necessary embankment and cutting. Basins are formed between the centre lock and those above and below it, the walls consisting of rough coursed stone laid in mortar, with a backing of clay-puddle, retained on the river side by a dry wall and a slope of stone and clay from the excavation. The combined lifts of the locks amount to twenty-five feet, the rock forming excellent floors. Over the lower piers of the upper lock is a rolling bridge, where the road from Brockville crosses, and continues over the river on the top of Merrick's old dam. A large block-house is built just above this bridge, and close to the lock, of fifty feet square, the lower story of rough coursed masonry, and the upper of timber, projecting over the basement: it is used as a house for the lockmaster and assistants. The total length of the canal is 1050 yards, and from a little above the head of the upper lock is an embankment two or three feet high, on the edge of the excavation, on the river side, to guard against the rise in spring-floods. The dam crosses the river a little below the entrance of the canal, and is connected with the embankment by a substantial rough stone wall, with a slope of gravel and broken stone in front. It is of wood, similar to that at Burrit's Rapids, and varying from six to ten feet high, as the rock dips across the river. The logs across several bays are constructed to be easily lifted, and the whole of the upper tier can also be removed if necessary. In front is a slope of gravel, and two or three steps of large stones, laid with care behind the wood work, break the fall of water, and protect the bed of the river.

At this place, now called Merricksville, there has sprung up, since the commencement of the work, a rather large village, and the roads to the frontier have been much improved. The whole clearance from Burritt's Rapids is known by the name of the " Lower Rideau Settlement," and after passing through the " Bush," about four miles on the road to Brockville, another clearance extends westward on a ridge of land to Maitland's Rapids, called the "Upper Rideau;" these were both settled principally by emigrants from the States.

Plate 2, fig. 7.—At *Maitland's Rapids*, a mile and a half above the mouth of the Irish Creek, and eight miles above the locks at Merrick's, a canal of about 450 yards in length is cut across a low swampy tongue of land, on the right bank of the river, at the extremity of which is a wooden dam, similar in construction to that described above. Its height is about seven feet, and an earthen embankment, nearly 400 yards in length, in which is a wooden regulating sluice, connects it with the wing-wall of the lock, which is crossed by a rolling bridge; the road from Brockville continuing on the top of this embankment, to a ford below the dam, which however is sometimes almost impassable, from the water backed up. The lock is situated nearly in the centre of the canal, and has a lift of only two feet three inches thrown on the upper gate, there being no masonry breastwork: both sills are of wood, bolted to the rock bottom, and the depth in the canal, above and below, is five feet six inches. An extensive swamp, crossed by a "corduroy bridge," about a mile to the southward, threatened to afford a new channel to the river when raised by the dam, and an embankment has consequently been formed across it, at the narrowest point that could be found.

Plate 3, fig. 8.—Four miles above is another single lock and dam, at *Edmons' Rapids*, also on the right bank. The lock is built close to the river, on a stiff clay foundation, the floor being composed of thick pine-plank, laid lengthwise, on sleepers which extend under the lock walls. The sill not being far advanced when the failure of the first trials at the entrance locks occurred, was laid in cement, as was also the case at nearly all the works up from Burritt's Rapids, and the stones notched to the course below. The canal, above the lock to its entrance into the river, is 215 yards, the excavation being all a very stiff clay mixed with boulders. The dam, which crosses the river forty yards above the wing-wall of the lock, is of stone laid on edge, founded on the rocky bed of the river, and instead of clay puddle at the back of the wall, the whole mass consists of broken stone and gravel. A cut-stone regulating sluice is formed in the right channel of the river, which was formerly separated by a small island, now crossed by the dam. Its height is about thirteen feet, and length of water-way 315 feet; an earthen embankment extends from the end to the wing-wall of the lock, being raised, as well as the abutments of the dam, above the level of the highest spring-floods. The lift of the lock is eight feet eight inches, allowing seven-feet water below, and five feet six inches on the upper sill, making the height of the masonry of the breast ten feet two inches above the wooden floor.

Plate 3, *fig.* 9.—At " *Old Sly Rapids*," one mile and three quarters distant, are two combined locks on the left bank of the river, founded on rock; the walls on the river side are built of an extra thickness of three feet of masonry, and faced with ashlar, instead of having any backing of earth behind them. The total lift is sixteen feet six inches, allowing five feet six inches water below, and seven feet on the upper sill. Both the stone sills were laid in cement, as at Edmons'. The rock floor of the upper lock having been much shaken by the blasting during the excavation, pieces of timber were shaped to fit the rock between the piers, and bolted to it. The dam, about 250 yards long, abutting on the wing-wall and pier of the upper lock, was constructed on the original plan on which it was commenced, but was raised, to prevent any water flowing over it, and a channel sixty feet wide cut through the rock, on the opposite side of the river, to the level of the bottom of the required navigation, in which was placed a wooden waste-weir with moveable logs as a regulating sluice, and an embankment carried from the dam to join it. Immediately below the locks, is a basin large enough for steam-boats to pass each other, and the extension below to the river, averaged eight or nine feet cutting, for a length of 450 yards, through a hard gritty sandstone.

Plate 3, *fig.* 10.—The work at *Smith's Falls*, three quarters of a mile above, consist of three combined and one detached lock, with a stone dam and waste-weir at the head of the first, and a small wooden waste-weir to raise the water to the level of the canal above the latter. The locks are all on the right bank of the river, and the excavation was through an irregular mass of rock full of fissures and springs. Upon this foundation the three locks are built; their combined lifts being twenty-five feet, with seven-feet water below, and five feet six inches on the sill of the upper; across the chamber-walls of the centre lock is a rolling bridge, and immediately above the locks a large basin is formed by embankments, extending on the south side 850 feet from the wing-wall, to meet the canal below the detached lock; and on the other, 180 feet to a cut-stone pier, which forms the abutment to a waste-weir constructed of timber, with four sluices to regulate the height of water, and extending 200 feet, to a rocky island, from the opposite side of which the stone dam crosses the river. This was built on the same plan as that at the work below, and was in some danger during its construction; its height is twenty-three feet, having been raised six feet above the required surface of the canal, to prevent the water flowing over it, as was at first intended. The depth of water in the basin, and in the canal, to the single lock, about 600 yards above, is five feet six inches, and the lift of this lock, which is built upon a solid rock foundation, is eight feet, allowing seven-feet water on the upper sill, which is retained to that level by a low wooden dam or waste-weir about four feet high, crossing the river nearly abreast of the upper-wing walls of the lock, from which on the south side an embankment extends, till it meets the high land, formed by two rough stone walls with puddle between them, and a slope of earth on each side, guarded of course against the rise of the river. The canal excavation extends about 300 yards above, and the same distance below this single lock.

A village has risen at this place since the first commencement of the canal, on the opposite side of the river; and as there are very good tracts of land, and several settlements in its rear, it will probably become of some comparative importance. Mills upon a large scale have been built by the first contractor for the work, behind the waste-weir of the stone dam; and they are connected with the houses on the opposite side by a wooden bridge crossing the rocky ravine, which was formerly the bed of the river, and down which still flows a quantity of water, that turns the abutments of the dam, through large fissures in the loose rock.

Plate 3, *fig.* 11.—A little more than two miles above *Smith's Falls*, is a canal with a single lock, also on the right bank, at what was called the " *First Rapids on the Rideau*," the lift of which, six feet four inches, brings the navigation to the average summer-level of the Rideau lake. The canal is a mile and a quarter in length, and the lock is situated about the centre; the excavation, above and below, being principally limestone rock. In the lock-pit it was through a ridge of stiff clay with boulders, from eighteen to twenty-three feet deep, and the floor is of hemlock plank laid upon sleepers of the same wood. The sills, and part of the breastwork, were laid in cement, as were most of the works above Long Island, and the walls have a guard of four feet eight inches over the surface of the seven-feet water in the canal. A lay-by for boats is excavated on the south side of the canal above, about half way between the lock and the river, and an embankment runs for a great part of the distance on the opposite side. At the head, the cutting is deeper, averaging about ten feet. Nearly abreast of the upper entrance of the canal, the dam crosses the river, formed like that at Maitland's Rapids, of a sill, upright, and brace of oak, the former bolted to the rock; with pine or hemlock logs across the intervals between the bays, the whole backed with steps of heavy

rough stone, four bays can be removed entirely, and the upper row of the whole dam, 365 feet long, to prevent any great rise of the lake in the spring. Cockburn creek, which enters the Rideau just abreast of the lock, rises, as has been already mentioned, near the Mississippi.

About four miles above this work, the river Tay empties itself into the lake on its north shore. A canal has been formed from Perth, eleven miles up this stream, by the inhabitants, to join the Rideau, calculated for steam-boats of twenty-feet beam, and drawing nearly four-feet water. The difference of level from the water below the town of Perth to the Rideau lake, was twenty-eight feet, and the different rapids that composed this fall were converted into pools of still-water by dams, on the same principle as the Rideau. The whole work consisted of six dams with waste-weirs, and five locks; of which four were of rubble masonry, and the fifth of wood. It was commenced in 1831, and finished in 1833, excepting the lower pair of gates of the lowest lock, which they could not manage to hang for some time. In a commercial point of view, this canal, though rough in its construction, and composed of not very durable materials, will, in connection with the Rideau, be of incalculable benefit to Perth and the country in its vicinity, as they had formerly no nearer market than Brockville, the road to which crossed a contracted part of the lake at " Oliver's Ferry," seven miles above the first Rapids. The length of the lake, from the last named works to the *Upper Narrows*, is twenty miles.

Plate 3, fig. 12.—The original channel is here closed by a wooden waste-weir, and a lock of four feet ten inches lift, built in a cut made through a narrow tongue of land. The foundation is a solid rock, which has been sunk three feet lower than was necessary. There is no breastwork, the lift being thrown on the upper gates, and both the sills are of oak bolted to the rock. On the north shore the land is bold and rocky, but on the opposite side an embankment has been carried from the waste-weir for some distance.

The object of raising this part of the lake, instead of keeping the natural level of the whole as the summit, was to save the expense and trouble of some very difficult rock excavation at the isthmus, between the Rideau and Mud lakes, the entrance to which is just three miles distant.

The length of this piece of canal is about a mile and a half, and a great portion of it is through a very difficult rock, partly granite : with the idea of saving a little excavation, the winding course of a gully, rather lower than the straight cut, was adopted, and the turns were afterwards found so abrupt, that many of the corners had to be cut away near the Mud lake. The cutting is very heavy, above twenty feet, and the lock is built close to the shore, on a rock foundation, without any breastwork ; the lift of eight feet being thrown on the upper gates ; a block-house is built close to the lock.   Had the Mud lake been raised to the same level as the Upper Rideau, by placing this lock as an addition to that at Chaffey's Mills, and the lift of four feet ten inches at the Narrows, added to that of six feet four inches at the First Rapids, the summit-level would have extended the whole distance between this place and Chaffey's Mills, about thirty miles, and the work at the Narrows have been reduced to merely deepening the old channel, or making a short cut across the tongue of land. Some low ground must of course have been flooded by this plan, both on the shores of the Lower Rideau and Mud lakes, but it is of little value, when compared with the saving it would have occasioned.

Descending into Mud lake, the route continues, through the narrow channel before alluded to, into Clear lake, and from thence enters Indian lake, through a cut of about 180 feet connecting them.

Plate 3, *fig*. 13.—At the outlet of this last, in a creek where originally were situated *Chaffey's Mills,* is a single lock of ten feet two inches lift, about five miles distant from that at the isthmus, seven-feet water is (at the average level of the lake) on the upper sill, and the same depth in the lock.   The floor is rock, and the material used for building, entirely sand-stone, obtained partly close to the works, and partly at a large quarry a few miles distant, which supplied most of the contractors for this part of the canal.   A channel is cut round the lock, to carry off the surplus water in the spring, with a waste-weir.   The lock-walls have a guard of six feet six inches over the surface of the lake.

Plate 3, *fig*. 14.—The creek below this work, winding between banks hardly above the level of the water, leads into Musquito, or Opinicon lake, and within a few hundred yards is the outlet into Sand lake, where, on the site formerly occupied by *Davis's Mills,* is a single lock, built of the same stone as that used at the last work, with a lift of nine feet nine inches between the surfaces of the two lakes ; seven-feet water being in the lock, and seven feet nine inches on the upper sill. A waste channel is cut round a rocky knoll on the left bank of the creek, where Davis's house originally stood, and the surplus water flows over a waste-weir into the lake below.   The embankment, from the head of the lock, to join the high land, consists of two rough stone walls three feet apart, with clay-puddle rammed between them, and backed on each side with earth and stone from the excavation. The lock-walls have a guard of five feet above the level of the lake. This work is two miles and a half distant from Chaffey's, and three miles across Sand lake, to the next in succession at "*Jones's Falls.*"

Plate 3, *fig.* 15. — The works at this station are perhaps the most striking of any on the whole line of communication, both from their wild situation, and their magnitude. The dam built across the ravine, down which the waters from all the small lakes above found an outlet, is sixty-one feet high, and about 130 yards long on the top, abutting on each side on the high rocky banks, consisting of sand-stone and a species of granite. The dam itself is built of the former stone, large smooth blocks of which are laid on edge, breaking joint all the way up : the thickness of this wall is about twelve feet at top, and the backing of stone and clay extends about sixty feet, with a slope of about five to one up stream. The whole base must be between three and four hundred feet.

By forming temporary sluices of rough masonry laid in mortar, alternately on each side of the dam at different heights, to carry off the water as it was raised by the progress of the works, the contractor, Mr. Redpath, who was luckily well qualified for the task, managed to raise this enormous mass to its height without any serious impediment. The water is turned by it down a ravine, about a quarter of a mile long on the right bank, opening a little above the dam. A narrow channel has been cut through the granite rock bounding this hollow, to a depth of fifteen feet below the raised surface of the lake, and a strong framed waste-weir, with sluices in the bottom, placed in it, its top on a level with the water. By this channel and the sluices, its depth is regulated, and the surplus finds its way into the old watercourse, a little below the dam. A single lock of fifteen feet two inches lift is placed about the middle of this ravine, its walls having a guard of five feet over the surface-water. A basin bounded by the high rocks, connects this with the three combined locks, of which the two upper have each a lift of fifteen feet, and the river-lock thirteen feet ; this supposes seven feet water to be retained in it ; and also seven feet on the upper sill of the detached lock, but only five feet in that of the upper of the three combined.

These locks are all built on invert arches, and of the same species of sand-stone used at the two last works. They have altogether a most beautiful appearance, and seem hitherto to have answered perfectly, notwithstanding their dangerously high lifts. More care was taken with the breastworks and sills, than with those of the other locks, and they have been secured with bolts and strong iron straps since their first trial. The gates also are necessarily of a stronger construction.

The navigation continues below these locks, through Cranberry lake, along a muddy creek, winding among a quantity of drowned swampy land, and across Cranberry marsh, entering the Cataroqui through a passage cut round the site of the old dam at the *Round Tail:* three quarters of a mile below which are the works at *Brewer's Upper Mills,* distant from Jones's Falls about eleven miles.

Plate 3, *fig.* 16.—The waters are kept up to their required level by a dam at this place, instead of that formerly situated at the Round Tail. It is of framed timber, similar to those described at some of the works on the Rideau, eighteen feet high, and backed with a quantity of large blocks of stones piled behind its whole height, with a slope of gravel or clay in front. The water does not flow over it as was intended, but is carried off by a sluice-way on the left of the dam. The two combined locks are on the right bank of the Cataroqui, at the end of a cut of nearly 400 yards, with a large basin immediately above them, in which the water is retained by a substantial earth embankment on the river side. They are built on a clay foundation, and floored with timber: the lift of the upper lock six feet, and that of the lower eleven feet six inches, the upper sill having one foot more water on it than in the lower lock at Jones's Falls. On the lower sill it is seven feet deep. The guard of the walls of the upper lock is four feet, which is necessary to provide against the rise in the spring, the Cataroqui being the outlet of a chain of small Lakes lying to the westward.

Plate 3, *fig.* 17.—At *Brewer's Lower Mills,* one mile and three-quarters distant, another wooden dam, about thirteen feet high, is constructed across the river, to retain sufficient water between these two works; and a single lock is built on the left bank, on a clay foundation, entered from a cut of about half a mile long, made through the low swampy land, to avoid an abrupt bend in the creek : the floor is of wood, and the lift of the lock thirteen feet two inches, allowing seven-feet water on both upper and lower sills, with a guard of three feet two inches on the walls.

From hence to *Kingston Mills,* ten miles and a half distant, the water is kept up to a level by the dam constructed at the latter place. The old course of the stream is followed for the greater part of the distance, with occasional cuts across its bends to shorten the distance ; but near Kingston Mills, the flat marshy country through which it flowed is completely inundated, and all traces of the old Cataroqui have vanished. The channel is here marked out through the mass of dead timber on each side, presenting the most desolate appearance.

Plate 3, fig. 18.—The dam at *Kingston Mills,* which retains the waters to this level, is constructed on the same principle as the others, of stone placed on edge. Its height is about thirty feet, and the length on the top, from the rock on which it abuts, near the wing-wall of the upper lock on the right bank, to the cut-stone pier of the sluice-way, which joins its abutment on the left, near 400 feet. There is a far larger mass of clay and broken stone at the back of the key-work, than was originally intended, with a very gradual slope up stream. An earthen embankment extends to the west, from the sluice-way, for about 1000 yards, till it meets the high land, and a similar mound of nearly the same length is also raised across some low land to the eastward, to prevent the water from turning the works. The locks, built of limestone, are four in number, the upper being detached, with a large basin between it and the other three, which are combined, and of an extra thickness on the river side, being without any backing and faced with ashlar. The lift of the detached lock is eleven feet eight inches, allowing seven feet eight inches water on the upper sill, and five feet in the basin, which is retained on the west by a large earthen embankment faced with a stone wall. A species of lay-by, or dock, is connected on the east side with this basin, large enough to receive a steam-boat, and with piers to which gates may be hung. The lifts of the other three are also eleven feet eight inches each, allowing eight-feet water in the river lock, making a total lift of forty-six feet eight inches from the surface of Lake Ontario to that of the Cataroqui, as raised by the present dam; and a long wooden bridge crosses the upper lock, the road to Montreal passing over it. The excavation was through a species of granite, and was an expensive as well as tedious undertaking, rendered still more so by the difficulty of procuring hands in the spring and autumn, owing to the very unhealthy situation. At all the works between the Isthmus and Kingston, as well as some on the other side of the Rideau lake, the same delay was experienced from this cause. At the head of the bay where the Rideau canal enters Lake Ontario, the depth of water is at all seasons sufficient, but within a mile of the locks a rocky shoal, as has been mentioned, crosses the route. A coffer-dam was at an early stage of the work formed round a narrow part of the channel purposed to be deepened; but the canal was completed, and in operation, before any thing was done to remove this impediment; so that steam-boats were obliged, at certain seasons, to unload at Kingston Mills, there not being four-feet water over this bar. A channel has however since been cut round it, without meeting with any rock that required blasting under water, and the communication is now uninterrupted to Kingston, five miles distant from the mills.

The size of the locks being the same throughout the whole line, their description has been deferred till now, as well as the general dimensions of the breadth and slopes of the dams and the excavations. The enlarged locks, as authorized by the committee in 1828, are calculated to pass a boat 108 feet long, clear of opening the gates, and thirty feet wide over the paddle-boxes. Their length between the pointed sills is 134 feet, and breadth between the upright piers thirty-three feet. The chamber-walls are ninety-five feet long, eight feet thick at bottom, and five feet at top (the batter varying of course with the height, which in combined locks has an awkward appearance.) Deviations have been made in some of the works in the breadth of the walls and piers : but any detailed account of these alterations and their causes, would be foreign to this report.

Such of the locks as are upon a rock foundation, required of course no floor. Those upon clay were generally built upon an inverted arch, but the floors of several last constructed are of hemlock or oak planking, laid upon sleepers of pine or hemlock, and are found to answer perfectly well. Piles were, in hardly any instances, used, and then only partially in front of the breast-work. The walls, including the ashlar facing, were laid in common mortar, and the joints pointed with cement, and cement grout subsequently forced into them in the manner already described.

The lower sills of the single locks, and of the sets of combined locks, are of oak, framed and planked, and where the foundation is rock, the floors of the recesses are planked also. In those which have no breastwork both sills are of wood.

In most of the locks, the stone sills only are of ashlar, cut to a mould; but at some of the works last completed, several courses of the breastwork below were cut, joggled, and laid in cement; precautions that ought to have been taken with all, and which would have saved the necessity of the long fox wedge-bolts and iron straps, subsequently found requisite to retain the sill-stones in their places. The upper sluices are placed in the centre of the culverts formed in the piers, and have of course a greater pressure to bear than if they had been at the upper opening on the pavement above. They turn on a horizontal axis, and are worked by a double chain passing round a crab placed over the man-hole. Originally they were of oak, but being found unequal to the pressure, cast-iron gates were substituted, which had afterwards to be strengthened with bars of wrought iron. The gates are of oak, planked with the same wood, or pine, and the scantling of the framework was the same for the high and low lifts. For the lower gates, and in the high lifts, they were found too weak, and have been since strengthened by additional braces. They were at first worked by a double chain passing round a crab on the pier above; one end of the chain was fixed to the front of the gate near to the mitre-post, it was then led along blocks, bolted for that purpose to the pavement, up the wall to the crab. After a couple of turns round the barrel of the crab, it was again passed down the wall along the same blocks on the pavement, round a pulley fixed to the sill, and then fastened to the back of the gate. This plan was so far convenient, that one crab answered the purpose of both opening and shutting the gates; but it has since been found so liable to get out of order, and in the case of the lower gates, so likely to cause delay and expense from a pebble or chip of wood getting entangled in the blocks, or chain, below water, that the whole system has been changed since the first year the canal came into operation. The lifts are in many cases higher than is general, or perhaps prudent; and in the connected locks, they are in some instances so unequal, that some trouble is experienced in managing the supply of water from one to another, as a lock of five-feet lift naturally will not supply sufficient water for one of ten. The invert arches can also hardly be said to act in the way their name implies; as the joints of the course of ashlar forming the floor, are only dressed a few inches, and of course the stones are not in sufficiently close contact. Luckily there is not a single instance of really a bad foundation throughout the line. The worst were perhaps at Hartwell's and the Hog's Back.

The dams were originally designed in all cases to act as waste-weirs, but the first experiment at the Black Rapids showed the impracticability of this plan, even in the case of one of the lowest dams built on the solid rocky bed of the river. They were, however, mostly commenced with this design, and the sluice-ways have generally been constructed in the opening left by the contractor to carry off the water during the erection of the dam; had they formed part of the original plan, they would in most cases have been differently placed, and the method adopted to regulate the height of water, better than the present clumsy one of raising and lowering squared logs thirty feet long, down grooves in the piers, which under a pressure of water, is a most troublesome operation. The shape of all the stone dams is a segment of a circle, whose radius is about equal to the chord of the arc, and the masonry is formed of stones of different lengths placed on edge, so as to break joint. This construction was adopted with a view to their acting as an arch, which can seldom be the case, owing to the want of solid abutments. The base of the stone-work, called in the contracts " arched key-work," varied according to the height of the dam. At the Hog's Back it was intended to be twenty-one feet; in dams from eight to twelve feet high, eight feet ; and all to have a slope of one-eighth their height. Clay-puddle is rammed behind this wall for a breadth of from five to eight feet, backed by a mass of earth up stream, with a base of about three to one (in some of the contracts an angle of twenty-three degrees), which however has been greatly increased in all the dams formed on this plan. Two of the last built (at Nicholson's and Edmons' Rapids), are rather different in their construction, and of a smaller section ; broken stone and gravel being substituted for clay, which has been found to answer much better ; and at the lowest lifts wood has been used instead of stone. The higher dams are decidedly too great an extension of a good principle : their work in building, the serious consequences which must result from any injury when built, and the number of connected locks they require to surmount their level, often more than balance the advantages of saving excavation, and the additional masonry and lock-gates. The excavation for such parts of the communication as really come under the denomination of a canal, is generally about twenty-eight feet wide at bottom, with a slope of two feet to one. In rock, however, the sides are nearly vertical, and where the cutting was trifling, and the line not straight, the breadth is in some places increased. Any detailed description of the machinery used in working, or building the locks, or of the more minute dimensions of the different parts of the work, would be out of place in this general account, which will close with a statement of the average prices of the different principal items in the contracts, and of labour and materials in Halifax currency. Plans and drawings of the locks and machinery were to have accompanied this Report, but they have been reserved till a description can be given of the alterations which have been made in them since their original construction.

The gates were made partly by carpenters employed in the workshops on day-work, and partly by contract at 100*l.* per pair, taking upper and lower gates together, the iron work and timber being supplied by Government, but the workmanship, fitting, and hanging, done by the contractor. None of this iron work was done by contract, but the blocks, crabs, and all castings were procured principally from founderies in the lower province.

Halifax Currency.

	£	s.	d.		£	s.	d.
Chopping and cleaning, per acre	4	0	0	to	4	10	0
Grubbing stumps, do.	15	0	0	—	16	0	0
Earth excavation under six feet	0	0	0	—	0	0	6
Ditto „ above six feet deep.........	0	0	9	—	0	1	0
Hardpans (clay and gravel)	0	0	0	—	0	1	6
Rock excavation, shelly limestone..........	0	3	6	—	0	4	0
Ditto „ granite, &c.	0	4	6	—	0	6	0
Puddling	0	0	0	—	0	1	6
Embankments	0	0	9	—	0	1	0
Dams, including " arched key-work," puddling, and whole section of earth................	0	4	0	—	0	6	0
Masonry of the locks, per cube foot..............	0	1	0	—	0	1	6
Oak timber squared, per cube foot, the price increasing with the scantling	0	0	4	—	0	1	0
Pine and hemlock, ditto	0	0	3	—	0	0	6

Per cubic yard. (bracket applies to Earth excavation through Dams rows)

The prices of labour were nearly as follows : —

	s.	d.	s.	d.	
Carpenters	5	0	to 5	6	
Masons	5	0	— 6	6	
Stone-cutters	6	0	— 7	0	principally by measurement.
Labourers	2	6	— 3	6	
Sawyers	5	0	— 6	0	
Smiths	5	0	— 6	0	

Of the total expense of the Rideau it is impossible to speak precisely, as many of the individuals whose property was flooded, or required for the purposes of the canal, received compensation during the progress of the work ; though the greater part of these claims are still under consideration, an officer having been appointed on the part of Government to investigate and report upon them.

	£	s.	d.

The amount expended up to the 31st December, 1830, appears, by the documents published for the information of the House of Commons, to be 715,408 15 6

And the estimates for the following year, to complete the works, was 88,365 10 0

Making a total of .... £ 803,774 5 6

This estimate includes 20,000*l.* for the purchase of land required for Government purposes, and 14,000*l.* for compensation to individuals. It is probable that this sum has fallen short of the whole expense of the communication, including the block-house for the defence of the most exposed stations, and, that taking the amount awarded for damages into consideration, and the alteration in the system of working the gates, &c. the whole cost of the canal will not be under 900,000*l.*

E. C. FROME,

Lieutenant, Royal Engineers.

Chatham,
   28th Feb. 1837.

Appendix B

Establishment of the Ordnance Observatory at Toronto

The following account was prepared by Lieutenant-Colonel Edward Sabine, R.E., F.R.S., as an introduction to his *Observations made at the Magnetical and Meteorological Observatory at Toronto in Canada*, printed by Order of Her Majesty's Government, under the superintendence of Lieut.-Colonel Edward Sabine, of the Royal Artillery, Vol. 1: 1840, 1841, 1842 (London: Published for Her Majesty's Stationery Office, by Longman, Brown, Green, and Longmans, 1845). It is the most comprehensive summary of the department's work in scientific research during the late 1830s and the 1840s. Sabine supervised these investigations. He was later regarded as one of Britain's foremost scientists (see *D.N.B.*).

INTRODUCTION.

THE importance of obtaining a correct knowledge of the elementary facts of terrestrial magnetism, for the purpose of supplying a foundation whereon the advancement of that science on inductive principles may be based, has of late years been strongly and extensively felt.

The geographical determination of the direction and intensity of the magnetic forces at different points of the earth's surface has been regarded as a worthy object of especial research, in journeys and voyages undertaken to remote parts of the globe, by continental philosophers of great eminence—by men, as Humboldt, Hansteen, and Erman, whose names are among the most distinguished in the age in which we live, for devoted and successful cultivation of the sciences most intimately connected with the physical history of our planet. In our own country the example thus set has been, at least zealously, followed by officers of Her Majesty's naval and military services, who, in the interval of a long peace, have given a portion of their time to such pursuits; and recently the large and liberal aid which the British Government has extended in the equipment of maritime expeditions, and in the promotion of magnetic surveys in parts of the earth which are beyond the reach of individual enterprise, gives a reasonable prospect that, with such assistance and such encouragement, maps of the magnetic elements, corresponding to the present epoch, and based on observations extended to almost every accessible portion of the globe, will shortly be completed.

But valuable as are the researches which lead to such a result, they accomplish but one part of the determinations required for magnetic theory. The *periodical variations* of the magnetic direction and force, and their comparison with meteorological variations also of a periodical character,—and those *secular changes* which, with slow but systematic progression, alter the whole aspect of the magnetic phenomena on the surface of the globe from one century to the next, and which, in their nature, are not improbably intimately connected with the causes of the magnetism of the globe itself,—present subjects of scientific inquiry not less important in the view of those who, by the inductive process, would seek to ascend to general laws and to the discovery of physical causes.

But for investigations of this class a laborious and sustained system of observation is indispensable; and establishments are required possessing an observing staff whose attention shall not be divided by other pursuits or other avocations. The absolute values of the magnetic elements,—*viz.*, of the Declination, the Inclination, and the Intensity of the Force,—and their periodical and secular variations, together with their mutual relations and dependencies, constitute the proper objects of research in a FIXED MAGNETIC OBSERVATORY. To these must also be added, as a distinct but connected branch, an investigation into the nature and laws of magnetic disturbances,—of those occasional and apparently irregular perturbations by which the magnetic elements have been found to be affected. These phenomena have attracted an extraordinary degree of interest since the rediscovery, in the present century, of their contemporaneous occurrence over large portions of the earth's surface;* and sanguine expectations have been entertained, that co-operative and simultaneous observation in different parts of the globe would lead to a knowledge of their cause, and that by their means we might possibly be conducted to a knowledge of the physical nature of the more stable forces engaged in the phenomena of terrestrial magnetism. For this branch of the inquiry, also, systematic observation is manifestly required, conducted on principles of uniformity in respect to times of observation and to instrumental means, and extended particularly

---

* The first discovery of this remarkable fact appears to have been made on the 5th of April 1741, by the contemporaneous and preconcerted observations of Celsius at Upsala, and Graham in London. (Hansteen Magnetismus der Erde, page 412, *et seq.*) Its rediscovery in the present century is due to a series of corresponding observations undertaken by Arago in Paris, and Kupffer in Kasan, in the years 1825 and 1826.

to those localities where the agency of the disturbing causes is most largely developed. In France and Russia, Germany and Italy, public establishments were formed for the purpose of aiding in the accomplishment of these objects; but, as yet, the part which Britain had taken was limited to the partial and desultory exertions of individual zeal, when, in 1836, the attention of British philosophers was specifically drawn,—by a letter from the Baron Alexander von Humboldt to His Royal Highness the Duke of Sussex, President of the Royal Society,—to the claims which magnetic science must be considered to have on a nation possessing such extensive dominions in all parts of the globe, and such unrivalled means of contributing to the advancement of the physical sciences, by the formation of suitable establishments in the localities in which researches might best be carried on.

The respect and consideration with which the Baron von Humboldt's letter was received in all parts of the United Kingdom, bear unquestionable testimony to the judgment of the illustrious individual by whom this appeal was adventured.

In the spring of 1837, the University of Dublin, at the instance of Dr. Lloyd, at that period Professor of Natural Philosophy in the University, voted the necessary funds for the establishment of an Observatory, in which all the researches connected with the sciences of terrestrial magnetism and meteorology might be systematically conducted; and, in the summer of the same year, on a representation to Government from the Board of Visitors of the Royal Observatory at Greenwich, a site was allotted for a Magnetic Observatory, to be placed under the general superintendence of the Astronomer Royal, and the building was commenced in the following year.*

From an early period of the meetings of the British Association for the Advancement of Science, the interests of terrestrial magnetism had received no inconsiderable share of the attention and exertions of its members. In the year 1834, a magnetic survey of the British Islands was commenced, and carried through in the two following years, by the joint labours of five of its members. This work, which though it did not require the aid of public funds, being gratuitously performed, may, with propriety, be regarded as a national work, was published in the annual volume of the Reports of the British Association for 1838, and has been followed by similar undertakings in other countries, conducted at the expense of their respective governments. In 1835, the Association called for a report from one of its members, on the state and progress of researches regarding the geographical distribution of the magnetic forces on the surface of the globe; proposing to ground on this preliminary examination an application to Government to aid in the prosecution of the inquiry in remote parts of the earth, unattainable by the means at the command of the Association itself, or of its individual

---

* At a later period a third Magnetic Observatory has been established in Britain, *viz.*, at Makerstoun, near Kelso, in Scotland. The expense of this Observatory is defrayed by the private munificence of General Sir Thomas Macdougall Brisbane, Bart., G.C.B. and G.C.H.

members. This Report, presented in 1837, was taken into consideration at the meeting of the Association at Newcastle in 1838, and a memorial was addressed to Government, which, being favourably received by Her Majesty's ministers, originated the naval Expedition equipped in the following year for a magnetic survey of the high latitudes of the southern hemisphere.

Deeming the opportunity a fitting one, the British Association availed itself of the same occasion, to solicit the attention of Her Majesty's Government to the expediency of extending the researches to be accomplished by fixed observatories, to certain stations of prominent magnetic interest within the limits of the British colonial dominion. The stations named were Canada and Van Diemen Island, as approximate to the points of the greatest intensity of the magnetic force in the northern and southern hemispheres; St. Helena, as approximate to the point of least intensity on the globe; and the Cape of Good Hope, as a station where the secular changes of the magnetic elements presented features of peculiar interest. It was also suggested, that the observations at these stations should include meteorological as well as magnetical phenomena; and a committee was appointed to present the memorial on the part of the Association, and to offer to Her Majesty's Government such explanations as might be desired.

Regarding it as not improbable that they might be requested to offer an opinion in respect to the agency by which the Colonial establishments might be effectively and economically conducted, and to the department of the State under whose general superintendence they might with propriety and advantage be placed, the Committee were naturally led to consider that the Ordnance department, and its military corps, by which the somewhat analogous service of the trigonometrical survey of Great Britain had been carried on for several years past with so much advantage and public satisfaction, combined effectiveness with economy in a higher degree than they were likely to be found united elsewhere. Having ascertained through one of the members of the Committee, himself an officer of Artillery,* that a suggestion of this nature would not only be regarded as unobjectionable by the then Master-General of the Ordnance, Sir Hussey Vivian (the late Lord Vivian), but that it would be highly approved by him,— as promising to add to the public usefulness of the Ordnance corps in time of peace, and thereby to their claims on the consideration of the country,—the Committee expressed it as their opinion, that the Colonial Observatories might be advantageously placed under the superintendence of the Master-General and Board of Ordnance.

The Government having signified a wish that applications involving such extensive arrangements, and a considerable amount of expenditure, should be strengthened by the concurrent support of the Royal Society, a deputation was appointed to express the cordial participation of that Society, in the recommendation both of the naval Expedition and of the fixed Observatories.

---

* The writer of these pages.

Her Majesty's Government having acceded to this joint representation from the two principal philosophical institutions of Great Britain, the preparation of the instruments for the naval Expedition and for the fixed Observatories, was undertaken by a Committee of the Royal Society, by whom also instructions were drawn up relating to the use of the instruments, and to the objects they were designed to accomplish. The instructions were published under the title of " Report of the Committee of Physics and Meteorology of the Royal Society, relative to the Observations to be made in the Antarctic Expedition, and in the Magnetic Observatories." The portion of the Report which related to the fixed Observatories was drawn up chiefly by Dr. Lloyd, and is written with his accustomed perspicuity and elegance, and with a theoretical accuracy which inspires the fullest confidence. In some few respects the instruments have been found inadequate, or the methods of observation unsuited, to accomplish the proposed objects with the required precision ; but when we recall to recollection the progressive improvement by which astronomical instruments have attained their present perfection, and astronomical observation its present precision, we shall not be surprised that magnetical science should have to pass through a somewhat analogous process ; and that especially, when the endeavour is first made to place it on the footing of an exact science, unforseen difficulties should present themselves, and practical inconveniences be found which had not been anticipated. The utmost attention has been paid at the Colonial Observatories to adhere closely to the instructions, except in the few instances where a departure from them was manifestly necessary. These instances are noticed in their proper places in the present volume, and have been for the most part discussed and improvements introduced in a second edition of that part of the instructions, revised by Dr. Lloyd, and published by the Royal Society in 1842. It is evident, therefore, that the date at which the Observatories should be regarded as fully effective for the different objects proposed in the Instructions of the Royal Society, must be taken at a later period than that of the commencement of the observations.

Whilst these arrangements were in preparation, it was proposed that as the Antarctic Expedition was designed to winter in Van Diemen Island in the two first years of its operations, the personal establishment of the Observatory at that station should be furnished by the Expedition itself, for the purpose of ensuring the fullest co-operation between the two branches of the magnetic service in that quarter of the globe. This proposal was acceded to by the Lords Commissioners of the Admiralty, leaving the other three Observatories to be provided for under the Ordnance department.

The trigonometrical survey, which for some years after its commencement had been carried on jointly by officers of the Engineers and Artillery, having in later years been executed by the Engineers alone, Sir Hussey Vivian was pleased to direct that the magnetic service under the Ordnance department should be performed by the officers and soldiers of the Royal Artillery exclusively. Accordingly a selection was made from the junior ranks of that corps by the Deputy Adjutant-General of the Royal Artillery, the late Sir Alexander Dickson, of officers who he deemed most competent to place in charge of the Observatories to be established in Canada, St. Helena, and the Cape of Good Hope; and each officer was directed to select three, afterwards increased to four, non-commissioned officers from the head-quarter establishment at Woolwich, as assistant observers. The officers thus selected were Lieut. (since Captain) Frederick Eardley Wilmot for the Cape of Good Hope, Lieut. John Henry Lefroy for St. Helena, and Lieut· Charles James Buchanan Riddell for Canada. By the instructions which these officers received from the Master-General of the Ordnance, they were directed to " carefully observe, and carry into execution, the instructions they would receive direct from the Master-General, or through Major Sabine; and to report to the Master-General monthly their proceedings in the execution of the duties they were employed on, transmitting their reports through the Deputy Adjutant-General."

The personal establishment of each Observatory was fixed at one officer, three* non-commissioned officers, and two gunners, one of the latter to act as an orderly, and the other as the officer's servant; the officers receiving the extra pay of staff-officers of their rank, and the non-commissioned officers and gunners a proportional increase of pay. A sum not exceeding 100*l.* a-year was allotted for each Observatory, to meet incidental expenses of all kinds, including the purchase of such new instruments as might from time to time be required. The total charge for each Observatory thus amounted to 392*l.* a-year. The officers and men stationed at the Observatories were on the spot in the respective colonies, should an emergency occur in which their military services might be required.** But, except in cases of emergency, their regimental duties were taken by the other officers and men of the regiment. In this point of view, therefore, the observations at the three Ordnance Observatories, recorded in this and succeeding volumes, may be regarded as a contribution to science rendered by the officers and soldiers of the Royal Artillery as a corps. In this light it was designed to be viewed by Sir Hussey Vivian; and it was so viewed by Sir Alexander Dickson, whose watchful care extended to this as well as to every other branch of public duty intrusted to the corps; and it is no less due to the memory of that most distinguished officer, than pleasing to record, that, by the superior influence of his station

---

* Increased in 1841 to four.

** An instance of this nature occurred at the Cape of Good Hope in June 1842, when one of the lieutenants of the small force of artillery stationed in the colony being killed in action with the insurgent Boors at Port Natal, Capt. Wilmot placed his military services at the disposal of the Governor, Sir George Napier, and was ordered by him to take the garrison duty of his corps at Cape Town until the occasion for his services had ceased. It happened fortunately that a few months before, Sir Hussey Vivian (without application, and possibly with a view to some such occasion arising), had sent an additional officer, Lieut. Clerk of the Royal Artillery, to be attached to the Cape Observatory for instruction in the use of magnetical instruments. In this case, consequently, no interruption of the service of the Observatory took place, as it remained under the charge of Lieut. Clerk in Capt. Wilmot's absence.

as the chief staff-officer of the Artillery, difficulties were frequently overcome which could not but occur in a service somewhat removed from the usual course, and facilities were obtained which have in many ways contributed to its efficient performance.

By the Master-General's direction, and with Dr. Lloyd's permission, the officers appointed to take charge of the Magnetic Observatories were ordered to Dublin, to receive instructions from Dr. Lloyd in the manipulation of the magnetic instruments in the Dublin Observatory, to which those prepared for the Colonial Observatories were similar. Dr. Lloyd's kindness farther extended to subsequent explanatory correspondence with the officers.

In the autumn of 1839, the several parties quitted England for their respective destinations—the detachments for St. Helena and the Cape of Good Hope in the ships of the Antarctic Expedition.

Early in 1841, in consequence of a communication from the Marquis of Northampton, President of the Royal Society, to the Master-General,—recommending that steps should be taken for the reduction and publication of the observations of the Ordnance Observatories, which had been regularly transmitted to the Deputy Adjutant-General in the shape of monthly reports,—Sir Hussey Vivian, having communicated with the Lords Commissioners of the Treasury and received their sanction, placed the reduction and publication of the observations under my superintendence, with the assistance of Lieut. Riddell (who had recently returned from Canada on account of health), and four military clerks, an office being allotted for the purpose in the Royal Military Repository at Woolwich.

In April of the same year, an application was made from the Lords Commissioners of the Admiralty to the Master-General and Board of Ordnance, with the sanction of the Treasury, proposing that the reduction and publication of the observations made at the Van Diemen Island Observatory, and by the Antarctic Expedition, should be comprehended in the arrangements made for the Ordnance Observatories: the Master-General and Board were pleased to assent to this proposal, and to direct accordingly.*

The publications consequent on these arrangements, previous to the present volume, are as follows:—

1st. In 1842, the magnetic observations made at sea by the Antarctic Expedition on its passage from England to Kerguelen Island, were reduced and printed in the Philo-

---

* In consequence of an application from the Royal Society, made in April 1839, when the Ordnance Observatories were in preparation, the Court of Directors of the East India Company ordered the establishment of Magnetic Observatories at Simla, Singapore, Madras, and Aden (subsequently changed to Bombay). These Observatories are conducted by officers of the East India Company's Engineers, and their observations, as they arrive in England, are transmitted to the Royal Society. No arrangements have yet been made for the reduction and publication of these observations; but it may be hoped that the Royal Society will shortly take steps for that purpose.

sophical Transactions, the expense of printing being in this instance defrayed by the Royal Society.

2nd. It appearing desirable that the observations made at the Observatories on days of unusual magnetic disturbance should be separated from those made daily at stated intervals, and published without delay for the purpose of comparison with each other and with similar observations made simultaneously in different parts of the globe,—the observations on such days made at the four Observatories, and by the Antarctic Expedition, in the years 1840 and 1841, were printed in a volume by themselves, at the expense of Her Majesty's Government, and published in 1843.

3rd. A continuation of the magnetic observations of the Antarctic Expedition, comprehending the observations made at sea in the first year of its operations within the antarctic circle, was also published in 1843, in the Philosophical Transactions, the expense of printing being defrayed as before by the Royal Society.

4th. " Magnetical instructions for the use of portable instruments adapted for magnetical surveys and portable observatories, and for the use of a set of small instruments for a fixed Magnetic Observatory, by Lieut. C. J. B. Riddell, R.A., F.R.S., Assistant-Superintendent of Ordnance Magnetic Observatories." This very useful work, which was much needed, was printed at the expense of Her Majesty's Government, and published in 1844.

5th. In the same year (1844), a third series of the magnetic observations of the Antarctic Expedition was reduced and printed in the Philosophical Transactions. This series contains the observations made in the second year of the operations of the Expedition within the antarctic circle.

In preparing the present volume for the press, the utmost attention has been paid to the condensation of the materials into the smallest compass within which they can be brought consistently with useful development and lucid arrangement. By printing the observations simply in the form of a journal, much labour and time would have been spared, but they would have occupied nearly three times the present bulk. In the portion of the volume appropriated to a discussion of the observations, conciseness and economy of space have also been much studied. In a work of such magnitude as that to which the present volume belongs, which is to include the results of the observations of four Observatories, an avoidance of all expense not strictly necessary seemed a public duty, even at the sacrifice, perhaps, in some measure, of that more handsome appearance which may with greater propriety be indulged in where the results of a single Observatory are concerned.

I gladly avail myself of this opportunity of expressing my sense of the valuable and cordial assistance which I have received from Lieut. Riddell, in the correspondence which is maintained with the Observatories, and in the preparation of the observations contained in this volume, as well as in training the non-commissioned officers to take part in the various processes of reduction which the observations require. His attention has been unwearied in the examination of the correctness of the coefficients, and in checking the reductions by every means which our small establishment will permit. In the remainder of the volume, his assistance has of course been of the most essential service in the description of the adjustments of the instruments at Toronto, which were chiefly made during the time that that Observatory was under his charge. And even in that part of the work which is necessarily most peculiarly my own, *viz.*, the discussion to which I have subjected the observations, I have found great pleasure and advantage in conversing with him over each sheet as it has passed through the press.

I have also great pleasure in expressing both Lieut. Riddell's satisfaction and my own with the non-commissioned officers who form the establishment of our office; their names are—

> Serjeant John M'Grath,
> Bombardier Samuel Hendley,
> Bombardier Francis O'Sullivan,
> Acting Bombardier Charles Organ.

Their diligence in acquiring a competency to execute duties which were of course wholly new to them, and the zeal and fidelity with which they have fulfilled these duties, deserve every praise.

PROCEEDINGS AT TORONTO IN THE ESTABLISHMENT OF THE OBSERVATORY

Leaving his detachment to embark with the instruments in a vessel bound direct to Quebec, Lieut. Riddell proceeded himself to Canada by the more expeditious route of the United States; and having waited on the Governor-General at Montreal to present a letter of introduction with which he had been furnished by the Master-General of the Ordnance, and communicated with the Commanding Engineer, to whom he was the bearer of instructions and authority to build an Observatory, he proceeded to examine different localities which were suggested as convenient sites. The preference was finally given to Toronto, in the then province of Upper Canada, where a grant of two and a-half acres of ground, belonging to the University of King's College, was offered by the Council of the University, with the sole condition, that the buildings to be erected should not be appropriated to any other purpose than that of an Observatory, and should revert to the college when the Observatory should be discontinued. The sanction of the Governor-General for the acceptance of this offer was received in January 1840, and the building was commenced as soon as the season permitted.

Whilst the building was in progress, Lieut. Riddell was permitted to make use of a small unoccupied barrack in the city of Toronto, as a temporary Observatory; and the instruments having arrived under charge of the detachment, they were placed in one of the rooms of the barrack suitably prepared for them, all the iron which admitted of removal having been taken away. The observations made in the temporary Observatory form part of the present volume. The buildings were completed and possession received in September 1840.

The Observatory is situated in latitude 43° 39′ 35″,* and longitude 79° 21′ 30″ west of Greenwich,† on a rising ground, about half a mile north of the city of Toronto, and 360 yards west of the University buildings. The height above the surface of Lake Ontario is 107·9 feet.

The following description of the Observatory is furnished by Lieut. Riddell :—

" The buildings consist of—

" 1. An Observatory having two apartments, one 50 feet by 20 for the instruments, the other 18 feet by 12 for an office or computing room, with a hall or vestibule 12 feet by 6. A small circular room for the transit theodolite is connected by a covered passage with the instrument room, and is placed at a sufficient distance from it to obtain a view of the lower culmination of some of the circumpolar stars.

" 2. A detached building, partly sunk in the ground with a view to obtain an uniform temperature, containing one room 18 by 12 for experimental determinations and observations of absolute intensity. It is situated about 80 feet from the Observatory, so that the instruments placed in it may neither affect nor be affected by the magnets in the Observatory. This building was erected in 1842, in conformity with the directions contained in the revised instructions of the Royal Society.

" 3. An Anemometer house, constructed so as to support the vane and pressure-plate of Osler's anemometer at a height exceeding 30 feet above the roof of the Observatory, and above the neighbouring trees.

" 4. A small shed for the Inclination circle.

" 5. Barracks for the officer and detachment.

" The whole of the ground granted by the college is enclosed by a picketing. The buildings numbered 1 to 4 are at the eastern end, within an inner enclosure. The barracks for the officer and party are at the western end.

---

* By circummeridional altitudes of the sun observed with a repeating reflecting circle, by Lieut. Lefroy, R.A.

† At all the Magnetic Observatories conforming to the instructions of the Royal Society, the time in which the magnetical and metereological phenomena are recorded is mean time, astronomical reckoning, at Göttingen ; the difference between the meridians of Göttingen and Toronto, has been taken as 5ʰ 57ᵐ 12·5ˢ, and is derived as follows :—

	h.	m.	s.
Göttingen, *east* of Greenwich, by the Nautical Almanac . . .	0	39	46·5
Toronto Observatory, *west* of Greenwich :			

	h.	m.	s.	
By 18 sets of M. C. Stars observed in 1840 . .	5	17	19	} 5 17 26
By Chronometric comparison with Boston, 1840 .	5	17	33	

	h.	m.	s.
Toronto, *west* of Göttingen	5	57	12·5

"The Observatory is built of twelve-inch logs, rough cast on the outside, and plastered on the inside, the laths being attached to battens projecting two inches from the logs so as to leave a stratum of air between the logs and plaster. The doors and windows are double, and the outer door has the further protection of a closed porch. The small room or office is provided with an open fire-place adapted for a wood fire; the instrument room has neither stove nor fire-place. No iron whatsoever was used in the structure, the nails being of copper, and the locks and other fastenings of brass. The instruments are supported by massive stone pillars, each formed of a single stone about six or seven feet long, imbedded in masonry to the depth of three feet."

The arrangement of the several instruments is shown in the accompanying plate, from drawings by Lieut. Younghusband, R.A., and includes the instruments recently supplied.

The personal establishment at Toronto consisted, at the commencement, of its director, Lieut. Riddell, with three non-commissioned officers, named in the order of their seniority—

> Corporal, now Serjeant, James Johnston,
> Bombardier, now Corporal, James Walker,
> Acting Bombardier Thomas Menzies.

London, 1845

Edward Sabine
Lieutenant Colonel, R.A.

Index